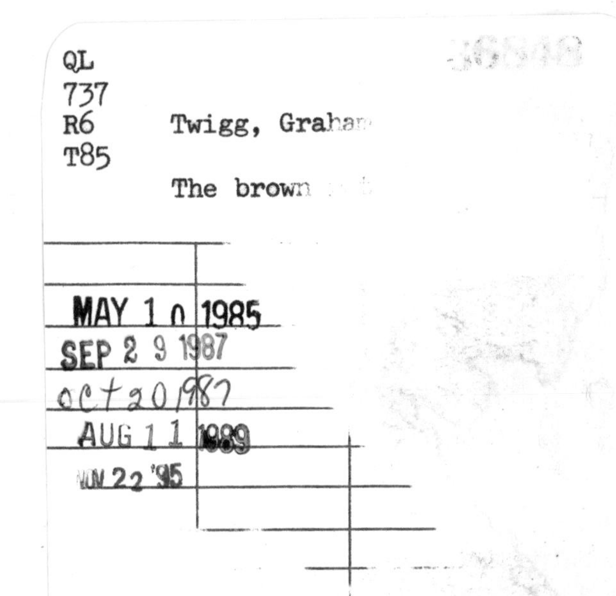

THE BROWN RAT

THE BROWN RAT

GRAHAM TWIGG

DAVID & CHARLES
NEWTON ABBOT LONDON
NORTH POMFRET (VT) VANCOUVER

ISBN 0 7153 6782 X
Library of Congress Catalog Card Number 74-20451

Set in 12 on 13pt Bembo and printed in Great Britain by Latimer Trend & Company Ltd Plymouth for David & Charles (Holdings) Limited
South Devon House Newton Abbot Devon

Published in the United States of America by David & Charles Inc North Pomfret Vermont 05053 USA

Published in Canada by Douglas David & Charles Limited
132 Philip Avenue North Vancouver BC

CONTENTS

LIST OF ILLUSTRATIONS

CHAPTER I

INTRODUCING THE RAT

When I asked a friend how he could best describe a rat, he replied: 'An animal with sharp teeth at one end and some nasty diseases at the other.' I thought at the time that that was rather neat, but now that I have to describe the rat I realise that, while it sums up the main reasons why the rat is especially important to man, it falls short of helping us to form a visual image of the animal.

Perhaps I am lucky, in that the rat probably needs less introduction than any other rodent – there can be few people who at some time or other have not seen one, however brief the encounter. On the one hand, they are large enough to be noticed and, on the other, they make their homes close to us, in our houses and buildings. To the countryman the rat is a fairly familiar sight popping in and out of holes in the hedgerow bottom or lopping along the lanes at night. His cat or dog is likely to do battle with them from time to time and he will no doubt suffer damage to his garden or to the potatoes he has in a sack in the garden shed. He will probably also be aware that rats are a potential source of disease to his pets and himself. Of all country dwellers, the farmer is the one most likely to be acquainted with rats at close quarters since a farm is, from a rat's point of view, the nearest thing to Utopia.

Rats are no strangers, either, to dwellers in poor city property. Even well-to-do town residents may spy a rat alongside a lake or canal, foraging on open spaces or around dustbins and in basements. Rats are probably less easily seen in urban surroundings, however, because most of them live beneath the streets or deep within the fabric of the buildings. Also, city dwellers have less intimate contact with their physical environment, outside of their own homes, than do countrymen.

Let us, then, try to get some clear idea of what constitutes a rat so that we can put it into its true perspective among other animals. Rats are, first and foremost, mammals. That is, they are warm-blooded, covered in fur and they suckle their young. Mammals come in many shapes and even more sizes, however, so the field must be narrowed still further. Within this assemblage of animals there is a series of subdivisions or orders. Each order groups together those mammals with certain common features; for example the order *Chiroptera* groups the bats and the order *Carnivora* the carnivores or flesh-eating mammals – the foxes, leopards, cats, dogs and others. By far the largest order, in terms both of actual numbers of animals and of numbers of species, is the order *Rodentia*, the rodents.

In order to tell a rodent apart from the other mammals one must first know something about the structure of teeth. One of the distinctive features of mammals is that their teeth are not all the same as are those of, say, a reptile (fig 1). In our own case we have at the front in both jaws a group of chisel-like teeth for cutting food – these are the incisors. Then, next to them, one at each corner top and bottom, are the pointed teeth, the canines. If you look at a dog or a cat you will see these very well developed. Moving back along the jaw are heavier-built teeth with hillocks on them: these are used for grinding the food small. There are two types, the pre-molars and the molars, but their differentiation is a matter for the expert.

Now, the supreme single characteristic of the rodent that

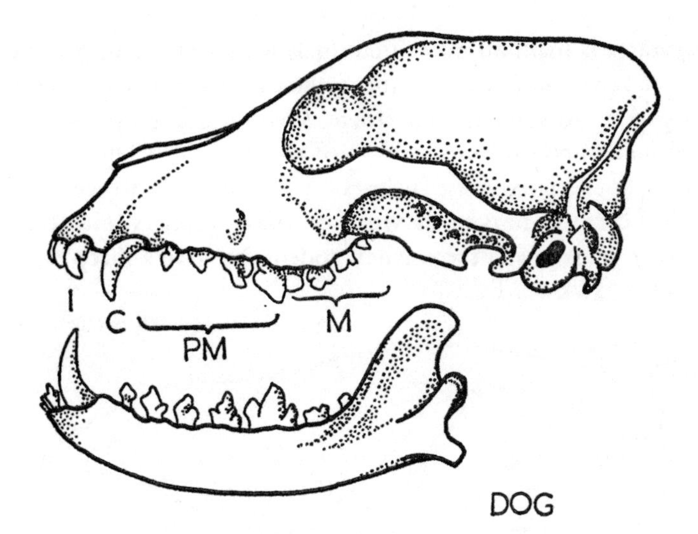

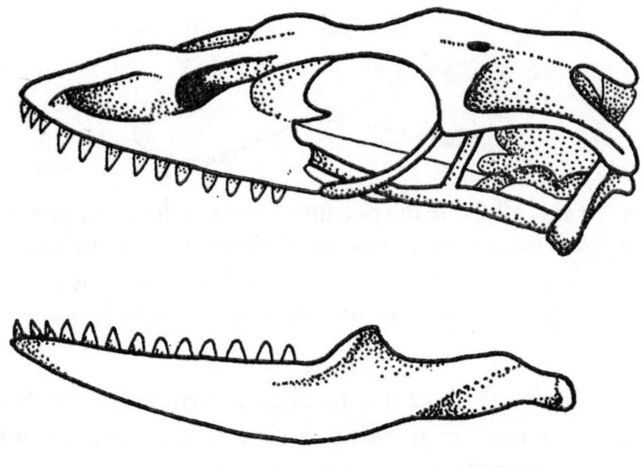

Fig 1 The teeth of a mammal (dog) and a reptile (lizard). Note the four types of mammalian teeth: the incisors (I), the canines (C), pre-molars (PM) and molars (M). The lizard teeth are all alike

separates it from all other mammals is that in the upper and lower jaws there is never more than a single pair of chisel-shaped incisors (fig 2). Since these do all the gnawing, they are a large and conspicuous feature of rodents. Walt Disney certainly appreciated these incisors – you have only to see some of his cartoon beavers at work or those gophers with their toothy grins! Even the smallest rodent can give a sharp nip if handled awkwardly.

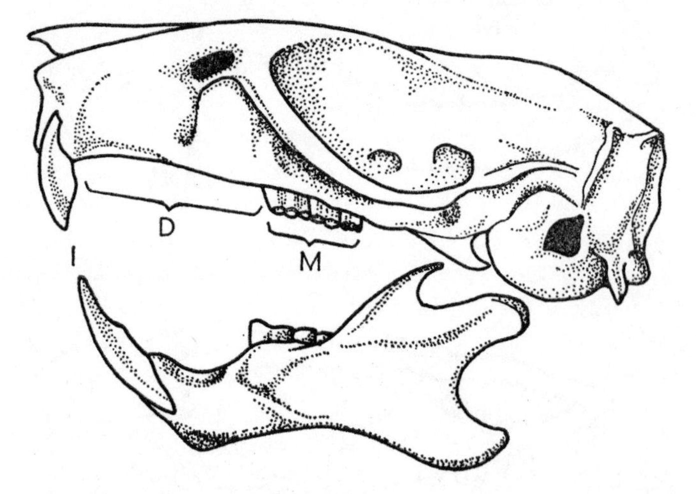

Fig 2 Rat skull. Note the two upper and two lower incisor teeth (I). There are neither canines nor pre-molars and the space they would occupy is known as the diastema (D). The remaining teeth are all molars (M), three top and bottom on each side

Rodents have no canine teeth and so there is a gap between the incisors and the grinding teeth. The gap serves a purpose: folds of skin can be inserted into it to close off the front part of the mouth, so that gnawed material is not necessarily swallowed. The ability to gnaw has resulted in many modifications of the teeth and of the skull as a whole, but there is no need to go any further for our purposes.

The largest rodent in the world lives in South America. It is called the capybara and tips the scales at 1cwt. The smallest rodents weigh about one-third of an ounce and several species of this size are found in many parts of the world. In between these two extremes are animals like the beaver and muskrat as well as the coypu, whose real home is in South America but which, thanks to escapes from fur farms, now does quite well living wild in the Norfolk Broads in Britain. It is the largest rodent in the British Isles and weighs up to 14lb.

The rat that most people are likely to see in Europe and North America is called, variously, the brown rat, the Norway rat or the common rat. This imprecision is unsatisfactory for the scientist, however. All living organisms, both plants and animals, are given a scientific classification name, based on the characteristics of the organism, which is the same all over the world. It has the advantage, therefore, that a biologist in Russia calls the organism by the same name as his counterpart in Peru, so that each knows what the other is talking about. All official names are in Latin and in the case of rats the term *Rattus* is used as a generic name to group the many species there are in the world today. The brown rat in this terminology is given the species name of *norvegicus*. Thus its full scientific name will be *Rattus norvegicus*. The black rat on the other hand is called *Rattus rattus*.

Rattus norvegicus is, by comparison with *Rattus rattus*, very much simpler to study. However, both species are complicated to some extent by the fact that within the species framework there are differences which are constant and which confer on the animals possessing them certain characteristics that justify our placing them in a separate category. The differences are not great enough, though, to justify placing them in a separate *species*, and so we call them sub-species. In order that we shall know to which species they are the sub-species they are given the normal species name, eg *Rattus norvegicus*, with a third name which is their sub-specific name. As an example, *Rattus norvegicus caraco* designates a sub-species of the brown rat from China. Black rat classification is very complicated – several sub-species have been described from all parts of the tropics, which is where they are most successful.

As you will see later, mention of the black rat is inevitable in any story concerning brown rats, and as the black rat is still found in some places in Europe and North America it might be as well to point out the differences and similarities in these two animals. Fig 3 and plate 1 show the black and brown rat and table 1 points out the chief differences between the two and for comparison gives details about the house mouse as well.

Biologists base their identifications of species on a variety of things; in the case of animals like rats and mice such factors as the length of the head and body relative to the tail length are of importance as is also the colour of the fur, the body weight and some body measurements. In experienced hands the sum total of all these measurements can lead to clear-cut identification, but care is needed, especially in certain environments or in certain populations at certain times.

Let us take, for example, colour: this is in any case one of the most difficult rat characteristics to study but it is of course the most obvious one and therefore worth discussing. If an inexperienced person assumed that the brown rat was always brown then, if he looked at a lot of brown rats from several localities he might, ignoring other characters and measurements, mistakenly identify many of them as black rats. The reason for this is that in brown rat populations there are always individuals who are black, or melanic, because of the deposition of the black pigment called melanin in the hair. Usually 1 or 2 per cent of a population is black but where the population is increasing rapidly, when food and cover is abundant, then as many as 20 per cent of the individuals may be black.

There have been few studies on melanism in wild brown rats. In the British Isles only 1·66 per cent of 1,266 rats from several localities were melanistic, a number which could not be accounted for by mutation or recombination factors alone.[1] A very interesting study of the topic took place in southwestern Georgia, USA, in the years 1954 to 1956.[2] In four counties the percentage of melanistic brown rats was 8·4, 9·3, 23·4 and 25·4. The two higher figures were from counties that represented

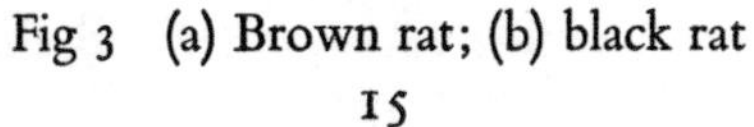

Fig 3 (a) Brown rat; (b) black rat

the front of a southward invasion by the brown rat advancing into territory formerly occupied solely by the black rat. It was suggested that the higher percentage of melanistic *Rattus norvegicus* at the edge of the advancing front might be explained by the behaviour and physiology that has been found associated with laboratory-reared black *Rattus norvegicus* mutants. These are less vicious and more docile than normal-coloured rats and are usually defeated in fights with wild brown rats. Thus it might be feasible to expect that as the *Rattus norvegicus* populations reached stability within the infiltrated regions, the normal brown rats would tend to push the black mutants towards the less populated areas where there was less competition.

Nor is black the only colour variant, for a research worker in Czechoslovakia[3] found that, out of eighty-three rats caught in a poultry farm in Central Bohemia, nine animals, seven males and two females, were of a strikingly yellow coloration. These were the only cases out of 3,023 rats which they had caught and examined in that area of the country. The different coloration was visible especially where dark hairs were absent, and was presumably due to a defect of pigment production in those skin cells that produce the colour.

Of course, if one takes all the other features into consideration there should be little difficulty in distinguishing brown rat from black, but care is needed.

As can be seen from table 1, the brown rat is bigger and heavier than the black rat. The coat of the brown rat lacks the smooth texture of that of its black cousin and is more harsh and shaggy. The photographs (plate 1) bring this out well. Most of these points of difference may seem small, almost insignificant. To a mammalogist, particularly a rodent specialist, however, they are vital points of distinction separating two superficially similar but behaviourally and evolutionarily quite different species.

Rats, both black and brown, together with the house mouse

TABLE 1

Comparison of brown rat, black rat and house mouse

Feature	*Brown rat*	*Black rat*	*House mouse*
Muzzle	blunt	pointed	pointed
Ears	small and furry	large and translucent	relatively large
Eyes	—	brighter and more protruding	—
Coat colour	grey-brown with a paler belly (black or fawn)	black, grey-brown or tawny and may have white or cream belly. Above it is usually darker along the dorsal line	grey-brown
Tail	thicker and shorter than length of head and body in cold climates. Longer than head and body in tropics. Scaly	thin and longer than head and body	
Weight	12oz	8oz	less than 1oz
In UK	everywhere in town and country: usually man-made situations including sewers	mainly ports but also in some large inland towns, especially those linked by rivers and canals with ports	everywhere
In tropics	mainly in ports	the common rat of countryside and town	town and country

(*Mus musculus*), may be regarded quite justifiably as serious pests on a world-wide basis. In some places such as Northern Europe the black rat is restricted to life indoors while the brown rat lives both indoors and out, as does the house mouse. Some mammals are classed as pests because they eat crops, others because they are the transmitters of disease to man or domesticated animals. The brown rat, and for that matter the black rat and house mouse too, score on both accounts and in addition do a considerable amount of damage to man-made installations. It is little wonder, then, that a lot of research is devoted to ways of, if not actually eradicating rats, then controlling them so that they shall do the least possible damage.

It is very easy to draw up a debit account for the brown rat with an impressive list of points against it, but it is of enormous value to research workers in all aspects of medicine and biological science. The albino strain of the brown rat is a docile animal which is bred in laboratories all over the world. Many drugs used to cure human ailments owe their existence to the fact that at some stage in their development their effects and side effects were studied by trying them out on white rats. No other animal has been of such use to man in this respect, and whatever we may feel about the wild variety we owe a debt of gratitude to the tame one. Its success as a laboratory animal is due to the fact that it breeds large litters so that useful numbers of genetically homogeneous experimental animals are available; also, it is a large enough animal to be used for surgical procedures, a feature that debars the much smaller laboratory mouse from much of the experimental work.

The brown rat has shown itself to be a highly successful species, capable of living a truly feral existence yet equally able to capitalise on man-made environments. Few, if any, mammalian species can match this versatility.

CHAPTER 2

DISTRIBUTION AND SPREAD

The brown rat has only five sub-species described; four of them come from China and the fifth from Celebes. The brown rat stock is thought to have originated in China and Siberia. It has spread outside Asia within the last 250 years or so and is now found throughout most of the temperate parts of the world, where it thrives, in the tropics, where it is confined chiefly to towns, and to the less hospitable parts of the world such as South Georgia in the Antarctic, Spitzbergen, the Aleutians and Alaska. It is surprising that even in comparatively recent times there should be reports of the spread of the rat in Russia and Siberia, for one is accustomed to think that since the species originated from that land mass it must have a wide distribution there. For example, rats appeared on the Trans-Siberian railway line only when the line was inundated with traffic during the Russo-Japanese war, that is ten years after it was built.

Because it is so well established today in the UK it seems hard to believe that no brown rat had set foot on British soil until around 1720, thereby achieving one of the most remarkable and successful invasions and spreading with considerable speed to all parts. But like most invaders it had to contend with some resistance, for there was already a rat there which was very well established. This rat, the black rat, was referred to in those days in a proprietorial manner as the old English black rat. It was black or dusky but English only by adoption, for its ancestral home was in Asia, probably India, although it had been in the British Isles a long time. We do not know when it arrived or who brought it but it had to come by sea, at least from the continent of Europe, since the British Isles had long been separated from Europe. The black rat spread, probably with trade, to Asia Minor and from there into North Africa and southern Europe, from which latter region the black rat of northern Europe developed. Although the black rat does not seem to have been recorded with any certainty anywhere in western Europe before the twelfth or thirteenth century AD,[1] there are reports of it being found in ancient sites in Switzerland.

It is possible that the black rat was imported into Europe by the navies of the Crusaders: although there is no clear evidence of its presence prior to the Crusades of 1095, 1147 and 1191, it was firmly established in western Europe shortly afterwards.[2] Yet there is no positive evidence that the rats came in with the Crusaders. If they did it seems curious that no indications of their importation have been handed down in legend or folk names linking rats with these warriors. Nor do contemporary chroniclers seem to have made anything of it. Certainly the black rat was known in twelfth-century Britain. Giraldus Cambrensis recorded large numbers of what he called 'mures' in Ireland and refers to 'the larger mice that are commonly called rats'. Five years after this a Welshman was eaten by rats, so it seems that they were established in late twelfth-century Britain.

The lack of reference to the arrival of the black rat with the Crusaders could mean that it was already well established in Britain by then. If we could be certain that the serious outbreak of a disease that was called the plague of Cadwalader's time in the sixth century AD was in reality bubonic plague, then the presence of rats would be practically confirmed, because wherever plague has turned up in the world in the last twenty centuries it has been brought there by rats. Before the Middle Ages, however, many epidemics were called plague,[3] and thus there can be no surety that this was bubonic plague.

There are no definite records of rats in Scotland till the sixteenth century. However, plague occurred north of the Border in the autumn of 1349 during the great pandemic that ravaged Europe.[4] There is more certainty that this was true plague since it was known to be in Britain then, and so we can feel more sure that black rats were in Scotland in the fourteenth century.

Further references to rats in the Middle Ages occur in Chaucer.[5] The harassed householder went to the apothecary for 'som poison that he might his ratouns quell'. Piers Plowman also mentions rats.

Whatever the area covered by the black rat and whatever debate there might be about its arrival, there is agreement that throughout the Middle Ages in Europe this rat held sway in town and village. Presumably, from its predilection for warm places, it did not live out of doors except perhaps in the summer, but was mainly to be found in the fabric of buildings. Countrymen's abodes were poor and the rats must have found abundant hiding to their liking here as well as in the towns. Hygiene was rudimentary. In the country hut all slept together on the floor in the single room with the animals which were herded in for the night. People in those days dropped their bones and scraps on the floor so that these provided a useful source of nourishment for the scavengers. In the towns things were even worse. There is little wonder that when pestilence struck it was able to proceed with lightning speed.

There is reasonable evidence that the brown rat reached Britain in the early part of the eighteenth century. There was at about that time a great plague of brown rats in Central Asia.[6]

The first brown rats arrived in Denmark from Russian ships about 1716 and reached Ireland perhaps as early as 1722. By 1729 the Irish, never a people to under-dramatise, claimed that there were rats the size of cats and rabbits. It was said that a woman and child had been eaten by rats at Merrion.

Pennant[7] in the early 1760s said that the first brown rats to reach England did so 'about forty years ago', ie about 1720. They were supposed to have come in Russian ships from the Baltic. The name 'Norway rat' seems to have arisen at this time for there was some idea that they had come from Norway in timber ships, but there is no evidence that the brown rat was present in Norway until 1762. It is said to have reached Sweden about 1790, France about 1750, Switzerland about 1809, Spain 1800 and Italy 1750.

The brown rat was soon unpopular and its arrival was not far in time from that of George I in 1714, a personage equally unpopular with the Tories, Jacobites and Roman Catholics, who said that the brown rat and the monarch came close together. The father of the naturalist Charles Waterton was convinced that the rat 'actually came over in the same ship which conveyed the new dynasty to these shores'. It is not surprising, therefore, that to many it became known as the 'Hanoverian rat'.

By the latter half of the century the brown rat had spread throughout the British Isles. In 1776 it first made its appearance at Selkirk where it tunnelled to such an extent that people were perturbed in case the burrows should cause their houses to collapse. By 1762 it had reached Anglesey where it is said to have eaten the standing corn while men were reaping it.

Although there is, unfortunately, a dearth of accounts of black rat natural history in Britain, it seems, perhaps because of its climbing abilities and preference for a warm habitat, never to have lived in hedgerows, fields or corn stacks, as brown rats now do. The brown rat on the other hand is at home under these and less favourable conditions: it lives in many parts of the British Isles out of doors all the year round and in addition has occupied sewers and mines. It has not, either in the British Isles or anywhere else, occupied the loftier places which the black rat prefers. Today the black rat hangs on in some ports and a few inland towns that were formerly connected to ports by canals. We find it in high buildings such as warehouses, where often the brown rat will occupy the basements of the same buildings. As long ago as 1768 the official rat-catcher to Princess Amelia in London remarked that the black rat lived in ceilings but brown ones in sewers,[8] so it appears that sewers were colonised early in the spread. Commenting further on the differences between the habits of the brown newcomers and the established black rat the royal rat-catcher said:

> The black ones do not burrow and run into shores as the others do but chiefly lie in the ceilings and wainscots in houses, and in outhouses they lie under the ridge tiles, and behind the rafters, and run about the side-plates: but their numbers are greatly diminished to what they were formerly, not many of them being now left, for the Norway rats always drive them out and kill them wherever they can come at them; as a proof of which I was once exercising my employment at a gentleman's house, and when the night came that I appointed to catch, I set all my traps going as usual, and in the lower part of the house in the cellars I caught the Norway rats, but in the upper part of the house I took nothing but black rats. I then put them together into a great cage to keep them alive till the morning, that the gentleman might see them, when the Norway rats killed the black rats immediately and devoured them in my presence.

From the mainland the brown rats soon spread to neighbouring islands where they wreaked havoc with the sea-bird colonies and other native wild life. Most of the islands around Britain have been colonised.

Clearly, the brown rat was soon in command in Britain and it has remained so to this day. Its great success has been due to a variety of strong points in its constitution. First of all, it can produce large litters rapidly. It is a hardy animal and can put up with more cold and general unpleasantness than most small mammals, especially the black rat. It is not fussy over food – it will quite literally eat anything. And, finally, it can adapt remarkably well to all conditions. By not being highly specialised it is not confined to any particular habitat or food and can make the best of all possible worlds. It is a specialist in non-specialisation, and that is the keynote of its success.

With all this to contend with it is scarcely surprising that the black rat fared so badly. In county after county the brown rat succeeded in taking over, and by 1890 the black rat was thought to be virtually extinct in Britain, though it seems to have lasted inland in some parts of rural Cheshire and North Wales until the 1880s. It probably survived on a number of outlying islands such as Lundy, Benbecula and South Ronaldshay. A colony of black rats was discovered at Great Yarmouth in 1895 and, of course, they continued to exist in ports. The invasion of the brown rat was a success, however, and the victors were soon accepted as naturally as their predecessors had been.

From Europe it was a natural progression for the brown rat to go with shipping to the New World. There, once again, the black rat was already widespread and abundant throughout the settled portions of the colonies – the date of its arrival in America is not known but it might even have crossed with Columbus.

The brown rat is thought to have been introduced into North America, probably from England, at about the time of the beginning of the American Revolution in 1775. Over much of the country the brown rat followed closely behind the early settlers, and the process of colonisation was rapid except in the high, rocky, dry mountain regions where, presumably, food shortage inhibited its spread. North America is a vast area with a wide variety of climatic extremes and, not surprisingly, the infiltration into some states has taken place only relatively recently.[9] For example, the rat was first reported at Denver, Colorado, in 1886 and by 1907 was present in most of the larger towns of the state. In 1888 there were none in Utah but were reported to be abundant in Albuquerque, New Mexico, and were known to occur in Arizona.

Wyoming was the next to last state to be invaded, in about 1919. Montana was invaded probably around 1923 but the first good records are in 1926, when a few rats were found at Fort Benton where it was said that rats had been introduced in the early days before the railroads when freight for the mining camps was taken off the boats there. For a time they were numerous but almost disappeared when the boats stopped. It was also said that at that time rats gained a foothold in Helena but later disappeared, suggesting that they did not thrive in the climatic conditions in Montana. By 1947, however, rats were established in two areas of Montana:[10] in the central part and in the east.

Extension of the range does not appear to have been easy in Montana, however. The human population is sparse – in 1940 there were only 2·3 people per square mile in rural areas, and most habitations are confined to cities and valley bottoms, as a result of which the rats are confined there too. Despite this, and despite the fact that the environment appears to be marginal for rats in the state, by building up strong populations as they extend they are capable of spreading across the whole state.

More recently the extension of brown rat range in south-west Georgia has been studied. In the years 1946–52 a large-scale invasion of territory by the rat has been recognised and it is clearly shown that where the brown and black rats occupy adjacent or overlapping habitats the brown rat usually dominates the smaller black rat.[11] In south-west Georgia three counties, Brooks, Thomas and Grady, were studied carefully. All three are primarily agricultural with many small farms growing peanuts, cotton, tobacco, corn and other crops. Several small towns cater largely for the farm trade and with minor exceptions rat harbourage exists in homes, outbuildings, barns and corn cribs which are one to one and a half storeys high and of loose construction. Between 1946 and 1952 the dominance of the brown rat increased annually. For the six-year period the invasion represented an average advance of about twenty miles and an occupation of around 1,000 square miles.

In southern cities of the USA the two rats may coexist in overlapping habitats for long periods but such cases are usually in dwellings having several storeys where the brown rat occupies the ground floor and basements and the black the upper storeys. South-west Georgia did not then have much multi-storey housing, and the author of the study suggested that the two species will not coexist on the same premises where there is competition for harbourage as well as for food.

The arrival of rats on the Pacific coast has been recorded as taking place about 1850, presumably mostly in ships. In May 1887 the brown rat was common at Vancouver, New Westminster and Victoria but had not reached Chilliwack, 60 miles up the Fraser river. It was there by 1894 but as late as 1905 had not become abundant although it swarmed at the three large seaports and along the coast. In 1947 it was reported[12] that the brown rat had not gained a foothold at any point in the interior between the Cascade and Rocky Mountains. In the years 1940–3 the black rat was said to be the most common species at North Vancouver, even in the woods a mile back from the port. This seems odd, however, considering the predilection of the black rat for warmth: the woods near Vancouver would scarcely seem a suitable habitat for *Rattus rattus*.

Canada was invaded from the rat reservoir that was present in Minnesota and Dakota, the first record of rats in western Canada being around 1900 when they appeared on the Manitoba–Dakota border (they were presumably in eastern Canadian towns already). By 1910 they had moved north following the old settled areas and were established in southern Manitoba.[13] In the years 1910–14 Saskatchewan became invaded at two points and southern Manitoba had infestations as far north as the Assiniboine river. During the next four years to 1919 all the then inhabited area of Manitoba was infested and the rats moved north-west in Saskatchewan. By 1939 practically all of the settled parts of east Saskatchewan had become infested.

It is noteworthy that the older settled areas of Manitoba were the first to be infested and this pattern continued until the rats reached Winnipeg. Thereafter, the invasion followed the railways to the west; this does not indicate any change in the rat advance, merely that from Winnipeg to the west the older settlements were along the railways.

The effects of natural barriers are more clearly seen in North America where the large rivers slow down migrations and the high mountain ranges provide an inhospitable cold zone which is not crossed without human aid. In Canada, the thinly settled and uninhabitable areas are rat-free.

As late as 1948 rats had not been recorded in Alberta, and the discovery of plague in the Prairie gopher (*Citellus richardsoni* Sabine) sharpened medical interest towards keeping the province rat-free because of the danger that rats would bridge the gap and transfer plague from the gophers to the human population.

In South America the rat has obtained some footholds but these are chiefly in the ports, and this is the pattern throughout most of the tropical parts of the world. In the Pacific the native rats still hold sway but the brown rat is extending. The disturbances of World War II introduced rats to new islands, yet it is surprising to find that the brown rat is probably still absent from such places as Guadalcanal,[14] which was the scene of such extensive fighting and the coming and going of enormous quantities of shipping.

CHAPTER 3

HOW RATS LIVE

In whatever habitat we study a group of rats, whether it be in a coal mine or on a farm, there are certain things they do that are fairly constant. For example, they gather food in a particular manner, care for the young in a set pattern, make nests in a special way and so on. The food they eat will vary with availability, as will the choice of nesting materials from one habitat to another, but this hardly affects the way they go about doing these varied functions. Thus we can talk about their general behaviour patterns and either relate them to those of other rodents or point out modifications in the basic rat pattern from one habitat to another. This general behaviour which covers every aspect of way of life is the focal point of survival and adaptability.

The strong point about rats is that they are not great performers in any one sphere but are competent performers in most, in contrast, for example, to a rodent like the coypu. This is a beautiful swimmer with webbed feet and very much at home in the reed-fringed waterways, but because it has become specialised for an aquatic way of life, it is confined mainly to that habitat. We would be astonished to hear of it in corn ricks or on board ship, and such specialists are as a rule less talented in other ways. I have already said that the rat is specialised in non-specialisation. In everyday terms the rat is a jack-of-all-trades or, in cricketing terms, a good all-rounder.

So far as we can find out from experimental work rats are probably not very good at seeing. This does not mean that they grope about myopically but rather that their capacity to analyse the environment optically is not good. Smell is fairly certainly much more important than sight and they probably recognise their mate or fellow rats of the group mainly by this sense. Tunnel-dwelling animals such as the rat tend to use the sense of touch in finding their way about, their whiskers moving regularly backwards and forwards to touch the area around them.

Let us, then, get some idea of the way a rat goes about its daily activities. We will begin by looking at how they go about making a home.

BURROWING AND NESTING

Burrowing is, of course, feasible only where there is soil which is not on the one hand too solid and difficult, such as a very heavy compacted clay, or on the other so light and dry that it is always falling in. Most agricultural soils, whether growing sugar beet in Norfolk or sugar cane in Hawaii, have the sort of consistency that falls somewhere between the two and generally form good tunnelling material. Hedgerows in Britain and Ireland are favoured by rats: the soil is untouched and certainly not pressed down, the roots of hawthorn or hazel forming the flying arches of the nests and homes. It is dry and, for much of the year there is good cover round the exits so that predators cannot spot them. With food nearby it makes a desirable residence.

Burrows have parts in which food is stored, living areas in which the nests are sited and holes in which to hide. It is said that the living area of a pair may have three to five entrances which join up after 18in to $4\frac{1}{2}$ft and soon afterwards run into the lair. Rats may have emergency exits which are hidden from outside view by a very thin layer of earth and from which they can bolt at speed if necessary.

Another long, blind passage is found near the living area: this is very narrow, so that it can be made water-tight with the rats' bodies. This is supposed to aid the rats at time of sudden floods. This blind passage is the last refuge of the rats – they are often found in it if they have not bolted before a burrow is excavated. Frequently in a well-infested bank the whole place becomes a honeycomb and under these conditions many families are in residence although we have no idea of how the living accommodation is shared out.

Rats often colonise disused rabbit burrows and those of other mammals. If they have to make their own homes rats dig with the front paws and remove soil from the tunnels by turning round and, using each front leg alternately, pushing the soil before them. This ensures that they emerge from the burrow head first and can thus see any danger in the vicinity. To emerge backside first, although it might be easier, would be the rat equivalent of being caught with one's pants down.

NEST AREA AND NEST-MAKING

The nest is where the rats live and rear the young. Into it is brought a remarkable variety of nesting material. In some boxes which were used for poison baiting, rats soon began to nest and the following list shows what they considered suitable for nest-making and for food as well:[1] pieces of gristle, orange peel, a government form dealing with swine, empty shell of water snail, pieces of fish skin, stems and leaves of elder, dead young rat (slightly gnawed), knuckle bone, empty cigarette packet, piece of electric cable, sheep wool, remains of dung beetle, slices of bread and butter, many partly chewed earthworms, potatoes, willow leaves, leaves of plantain, lumps of suet, piece of ox stomach, head of a starling, 1,000g 'Dairy nuts' cattle cake brought in during one night, 27g cotton seed cake brought in during one night, general rubbish such as grass, stones, straw, sticks etc.

All this is brought in by mouth one way or another – hard stuff is carried between the teeth but soft materials are carried in the pouches, the coarser elements being shredded with the incisor teeth.

The nest is constructed by pushing all the collected material into a big heap with the front limbs. The rat then crawls into the centre of the heap and brings in from outside all the material lying about in front of the nest. It then stretches out over the rampart of the nest and, taking each piece in its mouth, lays it back into the nest across the shoulders. Like most rodents, rats have an instinctive ability to build nests and they are made only when it is cold, never when it is warm.

LARDER AND DINING ROOM

The place in which food is stored is similar to the living area but has no nesting material and rats do not live in it. It is generally near the living area but may be up to a hundred yards away from it. In this the rat can eat without too much disturbance and, should anything happen to disturb a meal, can bolt into the hiding hole, always near the eating place – a simple blind passage running obliquely downwards.

Many rodents gather in provisions. Some of these are well known to us at an early age – most children will tell you that squirrels collect nuts for the winter. (They do, but then forget where they've buried them!) The rat collects only for its immediate needs and does not store for future shortages. It cannot therefore be classed as a typical hoarder. People who have carried out experiments with rats have come to differing conclusions about food collection. One research worker concluded after a series of experiments that there was a 'hoarding

Plate 1 (*above*) Brown rats (*Rattus norvegicus*); (*below*) Black rats (*Rattus rattus*)

Plate 2 (*above*) Early signs of rat damage with ragged edges to rick top. Later, the rick begins to disintegrate until (*below*) destruction is complete

instinct' which was activated by a psychological deficiency.[2] Yet another worker claimed that hoarding and eating had the same motivation.[3,4] He reckoned that if a rat was in the condition to eat in its home cage then it was also in the mood to bring food to the home cage. He said that the rat only hoards because the home as well as the food deposited at the eating place possesses positive values, whereas the eating place itself has negative values. Therefore, the rat carries its food to a place that suits it. When there was no gradient between eating place and home there was no hoarding.

Baby rats begin to take supplementary food, that is food in addition to their mother's milk, at twelve to fourteen days and are fully weaned at three weeks. It is said that the young prefer to take morsels of food from their mother as she is about to eat and so learn which food the species takes without having first to try out many unknown foods. Furthermore, young brown rats are supposed to learn by the example of experienced older animals which foods to avoid. This is said to be the case especially with poisoned food; it has been often noted that after early success in poisoning campaigns many rat baits were completely ineffective, supposedly because the older animals, which had had unpleasant experiences, avoided the poison bait and their example prevented the young rats from eating it. In refutation of these ideas, however, Dr Barnett, an authority on brown rats and their behaviour, says:[5]

> Choice of food has sometimes been believed to be influenced socially. Most such beliefs are mere myths – for example the story that adult rats warn juniors of the dangers of poison bait. There is no evidence of true imitation or observational learning by rodents, nor any behaviour even approaching parental instruction. Imitation can, of course, be simulated by associative learning: the presence of other members of the species can become associated with the presence of food; or the return to a burrow of an individual with food can indicate to others (as a result of previous experience) that more food is available outside.

Rats in the wild eat an astonishing variety of things and if all accounts are correct have a variety of techniques for food-gathering. It is said that they exhibit a 'search-grasp', that is that they will filter mud or dirty water between the fingers in the hope of dredging out some solid, edible component. They generally lap up liquids with the tongue and will also take water from a bowl with the paw and lick it from the wet hand. They will catch and eat flies from windows or the glass of cages. Small mammals are eaten in the wild, and in experiments 70 per cent of wild rats killed house mice offered to them and did it more quickly as they grew accustomed to it.[6] In contrast only 12 per cent of laboratory rats would kill mice. Most of the rats killed by biting the spine. In the wild, rats catch frogs, reptiles and birds as well as small mammals. They also go in for cannibalism. There are accounts of rats attacking rabbits, sparrows and even fully grown ducks.[7] Killing is usually done by biting the back of the neck.

Some writers have suggested that rats will kill poultry, but under normal circumstances this is doubtful. Chicks may be taken but hens could be tough customers to deal with, and there is one record of a barnyard cock turning the tables on a rat and pecking it to death. Rats kill the young of various small wild mammals and an adult rat has been seen to attack a leveret twice its own size.

Rats living on islands where sea birds abound have taken to catching birds, but first they had to learn a new behaviour pattern and this put singular demands upon their adaptability. Usually when rats are faced with birds, especially the soaring types, they flee. This reaction had to be replaced by a more suitable response:[7]

> ... they did not compare with the manner of slinking and lying in wait of the cat, but they approached the bird so casually, gradually, as though by chance, that in human terms one might say that they acted as though the bird did not concern them until they suddenly sprang at it from the nearest point.

When rats attack they jump upon the victim and beat it with their legs, at the same time uttering shrill screams. Being strong and armed with those keen incisors, they have been known to injure seriously fully grown polecats, the spreading out of the lower incisors after biting causing wounds that bleed profusely.

Naturally, rats are sometimes on the receiving end. When confronted by a cat both experienced and inexperienced rats stiffen. When cornered they face the attacker and will remain thus until the aggressor moves in for the kill. Initially the basic tendency is to flee but as the enemy moves closer the aggressive nature comes to the fore until finally the rat goes over to the offensive and attacks the attacker. This turning of the tables may so startle the predator that it will recoil and the rat may then seize its opportunity to escape.

Although not particularly specialised for aquatic life, rats can swim well. They swim mainly, and sometimes solely, by paddling the hind legs alternately. The front limbs are stretched out under the chin and take no active part in the proceedings. Provided the water is not too cold, they are reputed to be able to swim for hours. They can also stay submerged for two minutes and dive over 10-metre (11yd) wide stretches. Nothing is known about their diving. They can first swim at eight days.

It is recorded that a rat will catch young eels in the water and bring them ashore like an otter. An eye-witness of such a feat, James Hardy of Gateshead-on-Tyne, said:[8]

> On February 24, taking a walk with a companion, as we went along the side of the mill race at Swalwell, near Newcastle-upon-Tyne, we noticed a common house-rat making its way close by the edge of the water among the coarse stones that formed the embankment. Curious to know what it could be doing there, we watched its progress downwards, until it reached the outlet of a drain, into which it had just turned, when it gave a sudden plunge, and as quickly reappeared in the stream with a

middling-sized eel in its mouth. It made for the edge, where it soon regained its footing; and this, from the steepness of the bank, was a matter of difficulty, which was much increased by the eel, which it had seized a little above the tail, and was struggling vigorously to get free. The desperate efforts of the eel rendered his footing so precarious, that, rather than have a plunge for nothing, he was reluctantly obliged to drop it into the water. His first action afterwards was to give himself a good shake, both to revive his spirits and to rid his coat from the effects of his morning dip; and then, as before, he resumed his 'contemplative recreation'.

CARE OF THE BODY

Although very young baby rats go through some of the motions of grooming that they will do later, these are incomplete. As early as ten days the co-ordination of movements is already fully developed and the young rats can in addition lick and comb their fur. Straight away the newly born show scratching movements, but the hind leg moves in the air without touching the body. By the sixth day it can scratch itself and by the twelfth day it can lick its toes. Females wash the young.

Adults groom themselves carefully. First of all they wash the head with the fore paws. Most long-legged rodents, such as the rat, wash the head with the hands but in those species that have short legs the head actively assists the process, exhibiting a great deal of ducking and bobbing to enable the short fore paws to get to all parts. After completing the head they can lick the arms, legs, stomach and flanks. All scratching is done by the hind legs, which can reach almost every part of the body. During the toilet any injured parts receive special attention. If any part of the body is strikingly altered, for example is necrotic, they will gnaw it, a process that presumably gets rid of the diseased tissue. If rats lose the skin of the tail they will amputate the tail and eat it.[9]

The nest is likewise kept clean and tidy. Rats, and all rodents for that matter, never urinate or defaecate in the nest. Those rodents that live below ground use side passages in the burrow system as lavatories. If by accident any droppings find their way into the nest they are treated like any wet material and are removed.

LIFE WITHIN THE COMMUNITY

Starting with a single family, the normal processes of breeding result in very large social communities arising. These may consist of many hundreds of rats and it is said that the young of a pair and their descendants remain together so that tribal communities exist whose members live in harmony. The territory they occupy is delimited by an invisible boundary which is fixed by the inhabitants, and from their territory they drive away any strange rats that may happen to approach it.

On the island of Wangerooge in the North Sea, hordes of rats two thousand strong were estimated.[10] Such a horde was separated by a 50-metre (55yd) strip of land from the next horde. This boundary zone was the scene of battles when rats from adjacent hordes met and disputed the land. Sometimes in these large communities quite definite groups can arise, some dominant to others, so that there is an order of rank between groups which can have serious consequences on individuals who have the misfortune to belong to subordinate groups.

In mice, on the other hand, relations are usually quite straightforward: one male is at the top and all the rest are of equal, but lower, rank. Under such a regime there can never be any promotion problems except when the boss dies and there then must follow the sort of scramble that goes on in any board room when the managing director goes.

An American research worker[11] kept rats in a large, free

enclosure which was divided up by barriers but with places where rats could slip through. In the centre of the scheme was the feeding place. Starting with two females, a colony of 221 animals had arisen and these had formed up into several colonies, according to locality, each with a varying social structure. Near the feeding place was a colony presided over by one strong male, a 'boss' animal who ruled with a firm hand and prevented disturbances from other rats. The females of his colony were very much more successful at rearing young than were those of the other colonies in which several males lived. The further from the centre a colony was situated, the worse it fared. The one on the periphery fared worst of all and comprised few healthy animals. In this colony the males were in the majority of twenty-two to nine and only a few grew to maturity. The type of colony and its fate was decided by the characters of its founders. Where these were weak, subordinate rats, the colony was likewise. Where it stemmed from aggressive, healthy rats the colony was vigorous and remained so for several generations.

Young rats had their social position established in the third or fourth month of life. Older animals of dominating local colonies subjected the young of other colonies to repeated blows with their paws and thus established in their minds the fact that they were not, and never would be, as good as their betters. They were henceforth of lowly status. The more often youngsters were thus treated, the lower became their rank and the worse they developed physically. Those rats that suffered most were the young of the lowest-ranking colonies on the fringe of the population: they had a long journey to make to food and both there and back were subjected to beatings in the inhabited districts through which they had to pass. It was hardly surprising that females living in peripheral groups gave birth to smaller offspring, which in turn suffered badly because, since weight is a deciding factor in combat, they were unlikely to give a good account of themselves.

Within territories females have their own small sub-territories in which they are chased by a large number of males when on heat. All the males mount a female yet there is little conflict between these males.

AGGRESSION

Man is the only animal to kill members of his own species. Although many animals fight, or at least come close to fighting, they settle things in a sensible way. Up to a point, and wild animals know the point, the use of aggression and fighting behaviour is of considerable value in the maintenance of the species. Good stock is determined by selection of the best males, which vanquish their weaker rivals and perpetuate the line.

Another valuable function served by aggression is the carving out of territories, a feature of many animal species. A territory in the biological sense of the word is any defended area occupied by one or more animals. The defenders will attack other members of the same species that may approach this area. This aggression is a vital biological process. The area of territory is occupied by just the number of animals that it can support. If unlimited newcomers were permitted to settle it would soon become untenable for all: food would dwindle and the physical condition of all the inhabitants would deteriorate. One function of aggression, then, is to prevent over-population. Furthermore, aggressiveness towards the young members of a population may be instrumental in making them leave the parental area to find new home sites, and this extends the range of the species.

In most species, males are as a rule the more aggressive sex. Hormones and probably undefined processes of the central nervous system create a basic readiness to fight. As an example of hormonal influence, male mice become aggressive when they are about thirty-six days old, the point at which sexual hormones begin to circulate. Likewise, young male rats become aggressive when the same process takes place and their fighting instincts cease if they are castrated. The injection of testosterone, the male hormone, will increase in them the desire to fight, a tendency that can be halted by stopping the injections.

In general, the aggressiveness of an animal is influenced by experiences it has had. For example, house mice that have been beaten regularly when young are not prepared to fight later, and the earlier in life they suffer the hammering the less ready they are to fight. High-ranking mice are aggressive, but if they are submitted to a series of defeats they will finally accept the lowest rank. Similarly, subordinate ranking mice, if allowed to triumph repeatedly, can attain high rank.

In rats, experience counts for a lot. Animals that are experienced in fighting and are of high rank are not affected by castration for they will fight after emasculation, and, indeed, keep their old rank. Youthful experience in battle determines the future rank of brown rats. The place where a rat happens to be also determines how it fights, and here we see a direct human parallel. Like ourselves, who fight most fiercely when our homes are threatened, so in rats the readiness to fight is strongest in defence of the family area. Usually, therefore, the inhabitant of the territory wins, even if it happens to be physically weaker than the invader whom it is repelling.

Rats, like most animals, can size up an opponent, and when the intending combatants have weighed each other up and come to the conclusion that the other is a tough match, then a sensible course of action follows. The test of strength begins with threatening attitudes which go on for some time. The antagonists circle slowly around each other, the fur bristles up on end and the back is arched in a manner reminiscent of a cat. To add weight to the threat they chatter their teeth and emit a long drawn-out whistling, barely audible to humans.

Moving close together, flank to flank, they begin to press one against the other, the object being to push the opponent away. Nothing else happens – no biting or scratching, just this long, steady pressure which can go on for several minutes. The attacking rat keeps on pushing and edges its opponent a little bit at a time. There is no sudden jerky motion because this might cause the opponent to react by biting. Quite suddenly, all is over – the loser will give up and with one leap quit its opponent's territory.

Sometimes the contest takes a different form, with both rats standing upright face to face like boxers and thumping each other with the fore paws. They thus test each other's strength and, as in the case of the pressing technique, reach a decision without having to go any further. In the unlikely event of neither feeling that he is the lesser man there will be a few sharp bites on each side and the loser will then make off. If he is stopped he will throw himself on his back in a submissive posture to disarm his aggressor. This testing of strength without actually coming to real blows is a common feature of animal disputes. It avoids killing and gives the weaker members the chance to retreat while there is still time.

Occasionally a group of rats will evolve a different behaviour pattern. For example, in one case some female rats developed the technique of swift assassination of other rats. Their method was to creep up to the unsuspecting victims until within springing distance when they would pounce. They bit the victim in the neck, frequently penetrating the carotid artery, and all was over in a few seconds.

FACT VS *FICTION*

So much for the facts about various aspects of rat life. Perhaps more fascinating than facts are the many stories that have been, and still are being told, some of which confer on the rat the most remarkable skills and powers of reasoning. Many of these relate to food collection, and the following selection from various sources makes interesting reading. One author[8] claimed that rats had:

> even been seen to climb up the stalks of hollyhocks, clear off several snails at a time, bring them down with one paw, like an armful, and run off with them on three legs to a hole. On examination of the holes the inside was found to be strewn for some distance with broken snail shells.

It is quite likely that rats eat snails in some localities and that snail shells have been found inside rat holes, but in the above account I suspect that the finding of shells in the holes was the only piece of objective reporting in the matter and the gathering of the snails was invented to fill in the rest. A brown rat could not easily climb up, and certainly could not climb down, a hollyhock with great ease, and as for carrying snails under its arm *down* the plant and running off on *three* legs, that really takes some swallowing.

There are numerous stories of rats taking eggs and other awkwardly shaped objects. All these tales assume or imply a degree of *co-operation* between rats, by which is meant the assistance given by one rat to another in some task that could not easily be accomplished 'singlehanded'. Many and varied are the ways in which rats are alleged to engage in co-operative behaviour. The following account by Mr T. W. Kirk of the Colonial Museum, Wellington, New Zealand, is one such description:[8]

> I was standing in the doorway of a large shed, the further end of which had been partitioned off with bars to form a fowl-house, when I was attracted by a gnawing and scraping noise. Turning round I saw a rat run from a large dog-biscuit which was lying on the floor and pass through the bars. Being curious to watch if he would return, I kept quiet, and presently saw a well-grown specimen of the common rat come cautiously forward, and after

> nibbling for a short time at the biscuit, drag it toward the bars, which are only two inches apart, and would not allow the biscuit to pass. After several unsuccessful attempts he left it, and in about five minutes returned with another rat, rather smaller than himself. He then came through the bars, and, pushing his nose under the biscuit, gradually tipped it on edge, rat number two pulling vigorously from the other side; by this means they finally succeeded in getting a four-inch biscuit through a two-inch aperture.

Mr H. G. Hurrell of Devon collected together a series of accounts of rat co-operative activities which he published in the *Western Morning News*[12] and from which now follows:

> INGENUITY OF RATS
>
> How could a rat transport a fowl's egg up a wall without breaking it? A lady told me that this was a problem she once set out to solve. It was impossible to believe that an egg could be pushed by the rat's nose up the wall, although the wall could be negotiated by an unencumbered rat. Nor could the rat hold the egg in its forepaws and proceed on its hind legs, a method of transportation which has sometimes been witnessed. Humping along with the rat straddling the egg would clearly be impossible in this case. The astonishing method of rolling over and over like a hoop while clasping the egg with all four feet could never work up a wall. This method has been proved to be used by stoats, and once it is said to have been used by a rat. Only by waiting and watching did my informant find out how eggs were safely negotiated up the wall. She saw the rat take an egg under one arm and proceed up the wall on three legs.

Mr Hurrell goes on to describe the reports of four first-hand witnesses who have described how they have actually seen a rat on its back holding an egg being towed by another rat:

> Mr. A. James says: When I was a boy I saw a large rat towing a smaller rat, which lay on its back holding an egg, across the floor

of a granary near Plymouth. Mr. A. E. Abbot took me to an estate near Liskeard where he was formerly employed and showed me the now disused hen house where, he says: I have many times seen a rat lie on its side to clasp an egg with all four feet. I have watched another rat nose it on to its back by pushing the egg from one side and then take hold of its tail by the mouth and tow it away through a regularly used gap.

Mr. E. Harvey is a retired Metropolitan policeman. On his beat there was a store with a peephole coupled with a light left on to enable him to see if any unauthorised person was inside. He says: I have several times seen a rat in the store try to cope with an egg, without much success, until another rat helped by towing it backward, when it would twist itself on to its back, still holding the egg.

Mr. L. C. Faulkner has set down in writing what he has described to me. He writes: When I was a young policeman I was asked to investigate a suspected case of egg stealing at Rawden, Yorkshire. The hen house was a very old isolated barn with a loft. I concealed myself in the loft and watched through cracks in the floor. Four rats soon emerged and went to a nest with five eggs. One rat lay on its back holding an egg between its feet. The other three rats moved in front in line, each holding on to the tail of the rat in front. The rat with the egg was holding on with his teeth to the tail of the last rat in line, thus being towed along. They quickly disappeared down a hole and repeated the procedure until all five eggs had been removed. This was the only time in my police service that a detected crime did not call for police court action.

In these and many other instances the similarity of events might lead one to conclude that this type of behaviour is a normal everyday component of rat behaviour. Many biologists, however, doubt that this is so. There are also doubts about the accuracy of the observations in some cases, since rats operate mainly in darkness or twilight and these are not the best conditions for human observation without specialised equipment. Furthermore, in an animal whose whole way of life is geared to a series of not particularly intelligent responses to the various occurrences that beset him it seems doubly suspicious that he should use a degree of intelligence found only in monkeys and man in dealing with but one small aspect of his life and not carry over this ability to other spheres of activity.

I am not condemning all these reports out of hand. They are so plentiful and widespread that there could be more than a grain of truth in some of them. Nevertheless, the question could well do with someone taking a careful look at it, perhaps using some simple field experiments with wild rats in natural surroundings.

All biological processes are subject to change and development. That is what evolution is all about. Behavioural processes evolve like all other processes as conditions change and if an animal species is very effective it is justifiable to assume that its behaviour is capable of changing to meet the fresh demands placed upon it. The success of the brown rat so far seems a good enough guarantee that it will not be lacking in adaptability in the future.

CHAPTER 4

THE BREEDING CYCLE

Some reasons have already been given for the toughness and versatility of the brown rat. Another important point is that this animal, coming as it did from the harsh climate of Asia with its blistering heat in summer and searing cold in winter, was extraordinarily well equipped to cope with the milder climate of western Europe. Life out of doors in the Asian winter must have been exceedingly grim and the fact that the rat adapts so readily to man-made environments may indicate that he had already learned to do so in the bleak wooden townships of Asia.

The scene is set, then. An unobtrusively coloured, smallish rodent. Not so large that it is easily seen and hence hunted, but large enough to cover some distance if necessary and much more mobile than the very small rodents which live out their life span in quite a small area.

Another form of insurance for the future of the species, and this is very important to any animal, is the ability to produce enough offspring to make sure that there will always be plenty of your own kind to carry on when you are dead. All rodents do this very well, and it is one of the key pieces in the rat success story. The breeding cycle might in its narrowest sense be considered to involve just the physical production of young rats but I prefer to consider it as a total integrated process by considering the relationships of individuals of both sexes and the behavioural responses of young and parents to one another.

Although copulation and its resultant fertilisation is all that is strictly necessary to produce young and hence continue the species, the drive that stimulates this process is not allowed to go ahead at breakneck speed. There has to be, as in man, a progression from the first glimmerings of interest to the point of mating; a progression that must follow specific behaviour pathways and must have as its chief aim the overcoming of contact shyness that exists to a greater or lesser degree.

The amount of this contact shyness in a species is a measure of the way of life of the animals. It is particularly strong in solitary species and weaker in the gregarious species. Rats fall into the latter category, and as there is virtually no real contact shyness to overcome, courtship is more or less absent. The pattern of mounting and copulation is simple, especially in those animals that are high up in the hierarchy. When a female is in oestrus, or heat, she is pursued by all the males of the horde who in turn will mount her without any serious rank disturbances being evidenced by the pursuers. A pursuing male, when he catches the female, will first lick her genital region and then mount her. During this part of the proceedings the male rubs his fore paws rapidly against the female's sides. This elicits in her a posture called lordosis which is designed to make copulation possible. The whole business is soon over. After four to eight thrusts, during which ejaculation takes place, the female is released and both rats clean themselves.

Female capitulation is not always instantaneous. On occasion she will rebuff the advances of a pursuing male by biting him or, when he attempts to mount, by vigorously kicking with the hind legs. She may even run away: it is suggested that this is a leftover from a pattern of behaviour in a species in which strong contact shyness was evident, therefore indicating a former way of life in the species that was much more solitary than it is today.

There seems to be little to be learnt by young rats in their sexual attitudes and behaviour. If young females are brought up alone, when they come into their first heat period they will exhibit the characteristic copulation posture and will readily copulate in a short time. Likewise, males brought up alone show the male equivalent of the copulation posture and perform the sexual act like all experienced rats brought up in a group. Certain points in their behaviour have no doubt to be learned by experience – for example a small number of isolated males in experiments mounted correctly but failed to grasp the females from the right side.[1] This is soon corrected. The mount, the grasp, holding with the fore paws as well as the thrust are all carried out in the same way whether the participants are experienced or inexperienced rats. Isolated young females show a shorter heat period and lordosis than females that are experienced, however.

Keeping young males isolated is one thing, but bringing them up with other young males as their sole company is apparently quite another. Of those reared alone, 65 per cent copulated straight away, a figure presumably lower than that in animals brought up in the group, but in males reared only with other young males only one-quarter copulated.[2] It should be noted that these were albino rats. Homosexual tendencies develop in all-male rat communities.

The driving sex force in male rats is the male hormone testosterone. Castrated rats gradually lose their urge to copulate, although this can, of course, be revitalised by injections of the hormone.

Apart from obvious differences rats are, like ourselves, mammals and have their young in much the same way except that they have many more and they have them more often. The reproductive organs of the female rat consist of a pair of

Plate 3 (*above*) At 10 months, sugar cane is fine for rats but difficult for men to move around in when attempting to control them; (*below*) a cane field devastated by rats

Plate 4 (*above*) Good warehousing; (*below*) the worst possible way to store food

ovaries, one on each side not far from the kidneys, which produce the eggs or ova. These move into the uterus on each side and the uterus continues backwards to join in the mid-line and form the vagina which opens to the outside. Whereas a woman produces one and only occasionally more ova, the rat produces up to sixteen at a time. These eggs are fertilised by the male at copulation and the foetuses develop in the two uteri in much the same way as in human beings.

From fertilisation to birth is around twenty-four days. Not long before birth most female rodents increase their nest-building activities. The female appears more purposeful, scurrying back and forth with suitable material. Rats normally build sleeping nests except in warm conditions or climates, but the pregnant rat will be stimulated to build a nest in conditions where she would not normally build a sleeping nest. This nest-building activity is stimulated by hormones produced by the pregnant female.

The nest is very important for the welfare of the young in their early days. Mammals are warm-blooded: that is they possess the ability, by means of a series of complex internal mechanisms, to control body temperature so that it remains constant no matter what the external temperature does. In cold-blooded animals, by contrast, the body temperature goes up and down in roughly the same way as the air temperature does – hence they are sluggish when it is cold and are active in warm weather. Most mammals, and birds too, are active all the time. However, this ability to keep body temperature constant is lacking for the first days of young rats' (and of many other mammals') existence, so it is important to keep the young warm if they are to survive. If, for some reason or other, the mother rat cannot build a nest she will try to keep the young warm as well as she can with her own body, but she cannot do this effectively enough, and in most such cases her young will soon die.

As the time of birth approaches, the mother becomes increasingly restless until finally, as the process begins, she sits on her hind legs. In this position she can see the young as they emerge and as each is born, emerging head first, she takes it in her fore paws. Very quickly she removes the delicate membranes that cover the baby rat and eats them, along with the placenta or afterbirth. Each young rat is attached to its placenta by the umbilical cord through which, by means of blood vessels, it was nourished during foetal life. This is no longer needed and the mother bites through it and carefully washes the young. All this behaviour is innate – experienced mothers and those having their first litter behave in exactly the same way.

Behaviour patterns play an important part in the life of the rat and this is well seen in the responses of the mother to stimuli she gets from the young. For example, the mother eats the umbilicus down to its point of attachment to the body of the young. As a rule she will eat no further because, it is thought, the tearing and eating of the cord makes the young cry out and these cries inhibit the mother from going any further. Good evidence for this is shown when the mother gives birth to still-born young. These cannot cry and are usually eaten completely. Occasionally there are female rats that regularly eat their young, owing, apparently, to a disturbance in their make-up so that they do not respond to the protest cries of the young.

The mother, being heavy, could easily crush the nestlings to death but again there is a control mechanism. This operates only in suckling females, and prevents them from lying too heavily on the young. If non-suckling females are given a litter they will gather the young under them but will usually crush them to death.

Males take no direct part in caring for the young although they assist indirectly by preventing newcomers entering the territory and starting trouble. The presence of the male is tolerated by the nursing mother rat, but in some rodents that are solitary species the male is not welcome for he usually attacks the litter. Any attempt on his part to come too near is met by violent deterring action by the mother.

The mother rat not only has to provide milk for her young but has to keep a weather eye open for any sign of trouble that would threaten the litter. Any serious disturbance will result in her transferring the brood; if she happens to be suckling at the time she will make off and the young will hang on to the nipples and be dragged along. Some will fall off and be carried by mouth to the next refuge.

The nursing rat will add to her family young rodents of other species provided they are still at the suckling stage. But many adopted young die, either because the milk is unsuitable for them or because they cannot compete with the baby rats at suckling. If they do survive they can be accepted as part of the family and it is on record that young field mice (*Apodemus sylvaticus*) adopted by rats have lived with them for eighteen months.

It goes almost without saying that a female rat will defend her young not only from other species that approach too closely but from other rats. Pregnant rats, as they approach full term, are so aggressive to any rats that approach the nest that this ferocity is remembered and later enables them to leave the nest without fear of other rats going near it.

The period of maternal care when the young are taking milk is known as the lactation period and we say that the mother is lactating. Now, some mammals produce a family only once a year. These are usually large animals, which have a long gestation period and a long period of lactation so that most of the year is spent in rearing offspring. But a female rat does not even wait until the last litter is weaned before getting down to the business of having the next litter. Almost as soon as she has given birth she can mate again and become pregnant so that by the time the first litter is weaned the litter inside her is well advanced. Quite clearly this technique enables her to produce young at the maximum possible speed.

The way in which rats and other rodents can do this is because of certain changes taking place inside the females. These changes are cyclic in nature; in the unmated adult female there is a repeating sequence of events in the ovary, uterus and vagina resulting in a wave of egg production every five days. That part of the cycle when the female is sexually receptive to the male corresponds to the period of heat, or oestrus, in for example the dog or cow. If a female is successfully mated at oestrus the growth of the young in the uterus will take twenty-four days and this will interrupt the series of oestrous cycles. As soon as the litter is born the next oestrous period will take place and within a day or so of giving birth the mother will be once again pregnant while suckling the previous litter.

This rate of production will go on through the breeding season but not the whole year. In rats living out of doors in Britain and similar latitudes there is usually a period when breeding ceases (fig 4). This occurs in the winter months. During this period the oestrous cycles cease and this reproductively quiescent time is known as the anoestrous period. In the spring the ovaries, under the guidance of hormones from part of the brain, begin the cycle of egg production both in those rats that had already bred late in the previous year and in the young females which are now mature. Thus the next breeding season is under way. Let us now see what this ability to produce young quickly means in terms of increase of the rat population as a whole.

In the British Isles it has been shown that the average number in a litter at birth is between seven and eight.[3] Not all pregnant females will have the same number of young because fertility is closely related to body weight: pregnancies are more frequent among heavier animals and also the number of young per litter is larger. The fertility, therefore, is considerably higher in a population with a large proportion

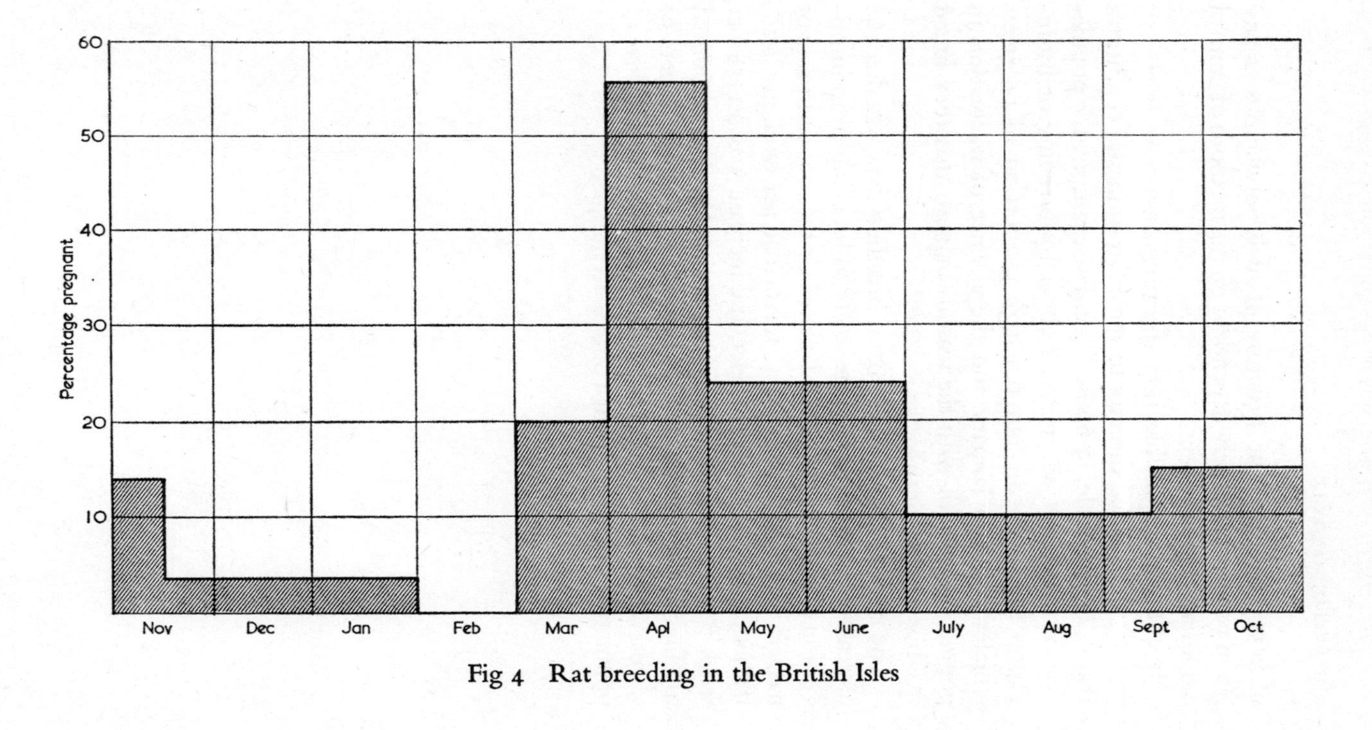

Fig 4 Rat breeding in the British Isles

of heavy animals. The structure of such populations at any given moment is largely dependent on the methods of control employed against them.

It also seems possible that the pregnancy rate (per cent pregnant) is related directly to the environmental conditions as it is, for example, in house mice. For example, the population of sewer rats will increase either by breeding or by invasion from outside, but the high proportion of pregnant females found in January and the absence of fluctuation in population growth with the seasons suggests that they breed steadily throughout the year.

The experts can calculate from breeding data whether the increase in a population is due solely to breeding or to immigration as well. It has been calculated that the inherent rate of natural increase for the brown rat is 0·1040 per week per head. If then, as in one sewer district, the increase was 0·1158 per week per head,[4] this rate of increase could not be accounted for merely by breeding. I shall talk about other aspects of sewer rat populations later on as well as the relevance of population increase to rat-control techniques.

CHAPTER 5

THE COUNTRY COUSIN

In general, the rat is a far greater problem in the country than in the town. It does most of its damage by either eating or spoiling food in the countryside and by transmitting disease to the countryman and his animals.

Although roads and buildings continue to multiply, an air trip will reveal that in fact most of the British Isles, and certainly most of North America, is still countryside. It is true that the majority of people now live in towns, but whereas the countryside lacks people on which rats can depend it is rich in both food and shelter and is as good or better a place for rats to live. In towns, rats are confined to certain niches. Underground they infest the sewers; they are also found in buildings, but these are separated from one another by hostile streets which are rarely crossed by rats in normal circumstances even at night. Human beings create an environment that tends to be noisy, well lit, and unpredictable in its capacity for change, all of which serve to make for a less stable life than the brown rat would prefer. For these reasons town rats are usually more concentrated than countryside rat populations, which tend to be spread more diffusely over quite large areas.

We can begin our study of the rat problem in rural areas

by finding out where rats live. The short answer to that is, just about everywhere! Not all places are equally favoured, however. Farmland, for example, is the best type of habitat. Old buildings on the farm are the most satisfactory for rats and the most difficult from which to expel them. Stone foundations without mortar provide a veritable honeycomb, and when hay is stacked on them the cover is absolute. Many farm buildings house bags of animal feed piled one on top of another behind which rats can live and on which they can feed by gnawing through the bags.

Progress towards better milk production has seen to it that the old cowsheds have to a great extent been superseded by cleaner, lighter, more efficient structures with plenty of concrete regularly hosed down and lacking the dark corners and holes once so prominent a part of the farm scene. Today it is rare to find corn kept unthreshed in stacks over the winter and so that particular source of large rat populations has been reduced. But modern farming, especially on large farms, has provided some alternatives. Where large numbers of stock or poultry are kept on deep litter all winter under one roof, the environment is good for rats, who are not slow to seize the opportunity – there is cover, warmth and food and a long period without risk of disturbance.

Piggeries have long been highly favoured rat haunts. Whereas in cowsheds, both old and new, the food is usually placed in wall racks, pigs are fed in floor troughs. Inevitably, there is spillage, both in doling out the food and by the pigs themselves as they nose the food about. Rats find this most acceptable and on occasion feed at the same time as the pigs. The fact that both can carry the same parasite may be an indication of the close association between the two over many years.

In the years following World War II, farms have tended to become larger and less varied, but farmland still contains a variety of crops all of which can provide food for rats at some stage of their growth. In the UK, for example, fields are often separated from one another by hedgerows, lanes with hedges on either side bisect the land and woods dotted here and there are used as rearing places for pheasants and partridges. Provided the soil is not waterlogged it will do well for a rat burrow and, if control is lax, signs of their occupancy will be evident in the banks around the fields and even in the fields among the crops.

Changing agricultural techniques can alter the pattern of rat communities, however. The widespread use of combine harvesters nowadays is one such example. Formerly, the cereal crops were cut first by scythe and later by the mechanical reaper, which tied the cut stalks into sheaves. These were stooked in the field and then built into stacks, more often than not in the farm stackyard adjacent to the farm buildings, although large farms stacked the grain in groups of stacks in or near the fields where it had grown.

All this took place in the late summer, but since the price of grain usually rose throughout the winter months many farmers left the stacks untouched until the following spring, thereby providing a perfect winter home and food supply for rats.

There was never any excuse for farmers not knowing they had rats in the ricks. The signs of damage were apparent quite early and the photographs (plate 2) show the stages after the brand new rick, through the early signs of damage and rat occupancy to the final decrepit stage of a heavy infestation. The pictures demonstrate that in the worst cases, damage was so severe that threshing was a waste of time and the rick a total loss. How much better to have settled for a lower price in the previous autumn.

By the outbreak of World War II it was clear that rat depredations on British food resources were considerable. A group of biologists assembled together at Oxford to study the effect that the brown rat, the house mouse and the rabbit were having on food supplies, both growing and stored, and to investigate control methods.[1] Corn ricks were one of the problems, and it soon became obvious that losses of grain in ricks due to rats and mice were heavy.

The ricks were usually colonised by rats before November; by then the average number of *active* rats in each rick was about sixteen but if one took the average total population, which would take into account baby rats not yet weaned, the number was about eighteen. Now, in the winter months, breeding of wild rats in hedgerows and woods slows down or stops altogether – it was found when this survey was being carried out that only 3 per cent of adult females were pregnant in these environments. But in the corn ricks 28 per cent of adult females were pregnant. This alone gives some measure of the satisfactory quality of corn ricks for rats. As a result, by the following April, which was the peak period of density in the ricks, the average number of *active* rats per rick was fifty-seven and the total population per rick was seventy-nine.

After April the rats began to leave the ricks, so that by June or July, when the last ricks were threshed, they contained, on an average, twenty-eight *active* rats each and an average total population of forty-two, a distinct fall from April but still many more than in November. As we know fairly accurately how much food a rat eats each day we can make some attempt at calculating the average quantity of grain eaten in a rick. Since the number of rats was not constant throughout the life of the rick we have to find some way of averaging out the rat population from autumn to threshing. When this is worked out and multiplied by the average daily amount of grain eaten by an average sized rat, we find that this typical average rick had no less than 7·03cwt of grain consumed by its rat population during the period.

And we must not forget that mice also inhabited ricks. The research workers who studied this problem pointed out that in ricks the mouse problem was every bit as serious as the rat problem and, furthermore, that there were large mouse populations in the ricks quite early in the season.

It was interesting that both rats and mice coexisted in relatively large numbers. Under such conditions there was a regional distribution of the nests with rats tending to concentrate in the upper half of the rick, and frequently, in the extreme base while mice tended to be found in the lower half of the stem of the rick.

At that time there was no doubt that the corn rick was a habitat of major importance in the ecology of rural rat populations. The inquiry showed clearly the great importance of early threshing, yet it was difficult to convince farmers that, far from gaining financially by holding on to ricks and threshing late, they were losing money. Of almost equal importance to early threshing was the necessity of killing all the rats found when the rick was dismantled. Legislation was passed making it compulsory to place a rat-proof fence around ricks that were being threshed in order to prevent the escape of rats into the fields where they would produce offspring which in turn would re-invade the next set of ricks.

Before modern rat-proofing techniques were expounded and every man had to work out his own solution, some farmers evolved their own methods of rick-proofing. Corn stacks and granaries were built on staddles (fig 5) and rat-proof piles. These consisted of stone piers surrounded by a wide, inverted pan- or saucer-shaped rim of tin sheeting. The piers were, ideally, three to three and a half feet high, with the protective rim near the top. Unfortunately, we have no knowledge of the effectiveness of these devices.

In the years following World War II, various attempts were made to proof ricks. Proofing materials present certain difficulties when it comes to keeping rats out; they must be smooth enough to prevent the animals climbing, made of material that will not deteriorate quickly when exposed to

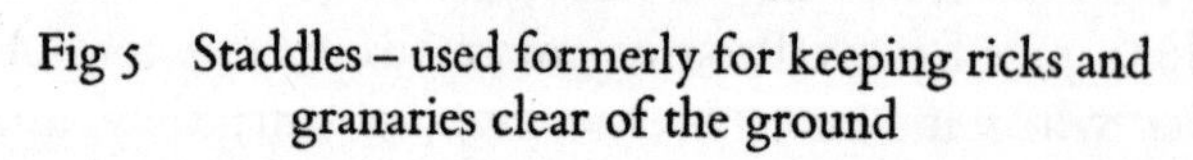

Fig 5 Staddles – used formerly for keeping ricks and granaries clear of the ground

rain and frost, be able to be fitted together so that there are no chinks and, in addition, must be buried in part to prevent tunnelling. And, of course, the material must not be chewable. Just about the only thing that fits the bill is chrome steel, which is steel that has been plated to prevent rust.

Some stacks were experimentally ringed with a chrome steel sheet 3ft high and dug down 18in below ground level: this was certainly effective at keeping rats out, although claw marks could be found 2ft from the ground and some frustrated digging attempts were evident. The cost of this prevention was high, however—steel of this quality was expensive and the digging was costly. Such a method was therefore hardly practical, and in any case combine harvesting was on the increase and its widespread acceptance has been a more effective rat-damage reducer than any amount of proofing would have been.

This loss in rick damage was, of course, only part of the farmers' losses owing to rats. Some, like the depredations on food stores, are obvious but others are not so clear cut. Take, for example, the feed put out for hens. In 1951[2] it was calculated, on the cost of foodstuffs then, that if a flock of 120 hens was supplied with dry mash, which the rats shared, the rats actually ate *more* than the hens at a cost to the farmer of about £1.50 ($4.00) per month.

This raises something of a dilemma for official rat-control methods. In terms of manpower costs, very small numbers of rats are costly to eradicate, yet if not cleared out they are the nucleus of the next population. It is therefore refreshing to find that the good old-fashioned relationship of cats versus rats has a part to play in this story. Several years ago, Charles Elton[3] investigated the role of cats in farm rat control and came to the conclusion that, while cats could not control a large, well-established rat population on the farm, they could, once the major part of the population had been poisoned, control the small number remaining as well as any rats coming in from the fields. Going back to the rats and hens situation the investigator there commented:[2]

> An interesting human problem arises, however, where the rats have been so reduced that further results are not sufficiently spectacular to retain the interest of the operator. Maintenance is tedious and costly in comparison with the number of rats killed. Cats, however, though unable to deal with well-established infestations, have proved admirable at preventing re-infestation after successful treatments with pre-baiting and poisoning.

In 1951 the Ministry of Agriculture, Fisheries and Food said[4] that on the question of rat damage in Britain: '... the recent statement of £25,000,000 per year is probably not an over-estimate'.

When talking about the cost of damage on the farm, or anywhere else for that matter, one of the problems has always been that of separating the cost of the container from that of the food contained. In the case of bags of cattle food or grain, the actual loss in the value of food *eaten* is often less than that of the cost of ruined bags and spilled or fouled food.

So far as control of rats goes, constant vigilance and prompt counter-measures are essential to prevent the buildup of large populations. It used to be the practice, and may still be in some parts, to have a day or two of rat-catching on the farm, usually in a slack season. This was not a day for guns, the armoury comprising terriers, ferrets and sticks. The technique consists of flushing out rats from their holes and nests by actual physical disturbance such as shifting corn bins and bags or putting a ferret into the hole. The terriers and men stand back and as rats bolt they are quickly snapped up by the terriers. Any rats that, by virtue of their speed and agility or because there are too few dogs, look like getting away are nobbled by the men. Surprisingly, a heavy stick is not necessary to kill a rat – provided you get it on the head a sharp rap with a light stick is sufficient because the roof of the rat skull is very thin.

It seems surprising, in view of the undoubted importance of the farm in rat ecology in the British Isles, that so few detailed studies have been carried out on the movements and biology of the rat in farms and on farmland. One reason for this may be that in order to carry out this sort of research a rat population must be allowed to increase to its maximum within the normal farming processes, and it is difficult to justify this in view of the damage that will occur due to the rats and, more seriously, because of the risk of persons on the farm contracting serious diseases from the rats. Thus here, as in other habitats such as the coal mines, research has had to be carried on as best it may while rat control was being practised.

In south-central Wisconsin a study was conducted on the wintering of field-living brown rats.[5] In summer the rat population was sporadically distributed at considerable distances from buildings, notably in grain fields, along the shores of lakes and streams, in maize fields, abandoned farm buildings and corn shocks. With the onset of cold weather all rats seemed to make for man's habitations; those that did not appeared to be localised about particularly good sources of food. They were seen to live along a lake shore quite late into the winter, living in holes in the banks and feeding upon dead fish, duck carcases and any available animal matter. In this climate the field-wintering rat population was never very dense, consisting in most instances of two to six rats occupying the corn shocks on a given farm. As the winter progressed the rats of observation areas usually diminished steadily in numbers until by spring there would be practically no survivors about the shocks. Indeed, except under exceptional circumstances the whole out of doors away from buildings and used dumps could be said to be virtually rat-less at the end of winter, and repopulation of such areas took place mainly through the spread of rats from buildings during the summer.

Owls and shortage of food were apparently the major problems facing field-living rats in winter – even at low population densities, they showed a peculiar vulnerability to the attacks of native owls. Clearly, then, in Wisconsin, which has a very cold winter, few if any rats survive out of doors and the farm must act as a survival point for wild rats. In Britain in most years the winters are not so harsh and rats can survive in the fields. Even so, it seems that the farms act as focal points where there is no doubt that the living is better than it is out of doors.

Although farmland supports many rats, there is an abundance of other places in the countryside where they can live and thrive. A favoured site is along the banks of rivers, especially when they are gently flowing and rich in fish and aquatic bird life. Large ponds and lakes with reed beds where water birds nest also provide a good living – there are eggs to be taken and perhaps young birds to eat as well as dead birds, fish and other animal life, to say nothing of a plentiful supply of green matter. Fish hatcheries invariably attract rats because on the one hand they can eat the young fish and on the other there is often to be found the spillage from the fish food. Since, in the countryside in Britain, burrows can be dug practically anywhere, the factor that controls the distribution of rats in localities other than farms is almost certainly food, and because the food supply from water courses tends to be fairly even all the year round, rats stick to rivers, streams and ponds.

Some waterside habitats prove to be lethal owing to flooding at certain times of the year. For example, in one bottomland west of Portsmouth, Ohio, the major roads were so elevated that the recurring high waters rarely covered them although the lands on either side of the road were flooded several times a year.[6] At the onset of flooding, brown rats were driven from the cornfields and were killed when attempting to cross the highways. In March 1938 an observer

Plate 5 (*above*) The drift leading to the underground workings; (*below*) sewermen at work. Hand abrasions are common and are probably the main portal of entry for leptospires in this occupation

Jack Black 1850

Her Majesty's Rat Catcher

Plate 6 (*left*) Jack Black, 1850 – Her Majesty's Rat Catcher; (*right*) recipe for destroying rats

RECIPE FOR DESTROYING

RATS.

RATS are fond of Epsom Salts—it sharpens their Teeth.

Put plenty down on the floor where they infest.

Epsom Salts makes them very ill and savage. Then they fight, kill, and eat each other.

RATS are Cannibals, and never migrate after eating Epsom Salts,

Epsom Salts are NOT Poisonous,

H. BRACEWELL,

95, CARDIGAN TERRACE, Leeds.

counted sixty-three dead rats in a one-mile section and the following day, eighty-four. At this time many more were killed by the local people in the water or taking sanctuary on piles of debris. The brown rat population of the town of Portsmouth, then around 50,000 persons, was controlled to some extent by movements of rats from the city to nearby cornfields in winter and their subsequent destruction with the arrival of the spring floods.

There are few rural habitats in which rats cannot find a living, even without the presence of man. Moorland is probably one of the most difficult and this was partly borne out in a survey of drift mines I carried out some years ago. I noticed that, almost without exception, those drift mines that were situated amidst moorland were untroubled by rats. The seashore, on the other hand, seemed to provide a respectable living. In 1960 David Drummond, of the Ministry of Agriculture, Fisheries and Food, studied the food of rats living in an area of sea wall, saltmarsh and mudflat on Bridgemarsh Island in the river Crouch in Essex.[7] Formerly farmed, it is today a Spartina grass-covered mud flat, the rats burrowing in the clay soil of the sea wall. These rats ate the tough Spartina grass all the year round and a good source of protein was provided by the shore crabs and sandhoppers. A few dead sea birds were consumed and, during the spring and summer, ground beetles were frequently eaten. Various dicotyledonous plants completed the main items of the diet. Since the rats readily eat Spartina and as rat traces in some parts of East Anglia are more common where Spartina grass grows, it was suggested that Spartina might affect rat distribution in some coastal areas in the way that cereals do on agricultural land.

So far in our study of rural rat economy it has looked as though rats have a free hand, living more or less where they please and eating everything edible that they can find. We must not lose sight of the fact that in the countryside the rat itself forms a considerable proportion of the diet of many animals. On the farm the rat can form part of the diet of cats. A study of the food of feral house cats in Oklahoma[8] showed that rodents made up 10 per cent of the diet of cats in the residential sector of a military reservation. Because this was a human settlement it is a fair bet that rats made up a good part of that 10 per cent.

In the British Isles the chief rural predators are the fox, stoat and weasel together with various owls. In the late 1960s, in a study in Ireland,[9] the rat made up 17 per cent of fox diet. An interesting point is that rabbits, but especially rats, taken by the fox were immature animals. Young rats will have less experience and perhaps when looking for their own residential territory are more easily picked off by prowling foxes. In Ireland the rat is important in the foxes' diet throughout the year but in late autumn there is a rise in its consumption. In studies on fox diet the rat is eaten much more frequently by Irish foxes than by their English counterparts. It is thought that feral rat populations may be relatively larger in Ireland where the smaller fields and the consequent increase in hedges and ditches could provide more cover.

Owls find rats to their liking and from the economy of hunting effort a rat, even an average-sized one, will provide at least five or six times as much food for the same effort as will a field mouse. The diet of an owl reflects the small mammal types in the vegetation over which it hunts—if it hunts over rubbish tips or farmland it will find more rats than if it frequents moorland pastures where short-tailed voles predominate. Similarly, on a larger scale from one geographical area to another, a change in the species of small mammals inhabiting the region will be seen in the owl's food. There are marked differences between the small mammal faunas of Ireland on the one hand and England, Scotland and Wales on the other. In Great Britain as a whole the barn owl preys on eleven species of small rodents and insectivores and at least six of these it takes commonly. In Ireland, though, only four of these are widespread and one of the four is the rat. Obviously the food of the barn owl must differ between the two islands and it does, to the extent that in Ireland the rat forms 30 per cent of the diet, a much higher proportion than in England.

The rat makes up 20 per cent of the food of Irish long-eared owls[10] but is apparently not equally available to the bird at all times of the year. In summer, 6 per cent of the food is rat but in the autumn rats are the main species taken, making up as much as 50 per cent of the prey. The reason for this is thought to be due to the rats' change of home between summer and winter. During the summer there is dense cover in the fields and surrounding vegetation but when autumn comes the crops are harvested and the vegetation begins to die down. Food and cover are scarcer and rats, temporarily unfamiliar with their surroundings, will be crossing strange terrain looking for food and heading for farm buildings and ricks. It is during this period that they will be especially vulnerable to owl attack. The amount of rats in the owl's diet is affected markedly by the proximity of human dwellings and water.

Other predators on rats are, of course, stoats and weasels, otters ranging up and down river banks, and a relatively new predator to Britain, the feral mink which favours water courses as its haunts. Nevertheless, even in the presence of this assembly of rat-hunters sizeable field rat populations can exist, especially in farmland where rats can be brought to a very low level only by the main predator, man, with his large-scale extermination methods.

Gamekeepers have always waged war on rats, and on a lot of other animals too; looking back through estate records in Britain one is amazed at the vast numbers of rats killed in years gone by compared with the kill today. For example, on one large estate in East Anglia keepers killed 14,662 brown rats in 1903. By 1915 when, presumably for war reasons, the game department was disbanded, the years' catch was steady at around 6,000 per annum. From 1919 to 1925 no records were kept but in 1926 over 10,000 rats were killed. Between then and 1942, acreage and keeper pressure remained constant and rats killed decreased to 1,500 per annum. In 1942 the number of gamekeepers was reduced and land was turned over to the plough and airfields so that comparisons are not easy. Nevertheless the rat kill remained at about the 1942 level until 1954. In that year myxomatosis wiped out very large numbers of rabbits and, presumably because the fox and some other predators had to look elsewhere for their supper, the rat kill decreased further. In 1955 it was 990 and has fallen steadily until 1966 when only 466 were killed. Despite a decrease in keepers and estate size the number of rats per acre has fallen dramatically in these sixty-three years and a rat density of that obviously present in 1903 would raise an outcry today.

As just one example of the large numbers that used to be present in certain places, the following account of a French slaughterhouse will suffice.[11] In 1840 there stood at Monfaucon a horse slaughterhouse which it was proposed should be moved to Paris. It was stated that the carcases of the slaughtered horses, which sometimes amounted to thirty-five a day, were cleared to the bone by rats in the course of the following night. This excited the attention of a Monsieur Dusaussois who made the following experiment; he placed the carcases of two or three horses in an enclosure which permitted the entrance of rats by certain known and closeable paths. Towards midnight he and some workmen entered the enclosure, closed the holes and in the course of that night killed 2,650 rats. He repeated the experiment and by the end of four days had killed 9,101 rats and by the end of a month 16,050 rats. During the process of these experiments other carcases were exposed in the neighbourhood so that in all probability Monsieur Dusaussois attracted to his enclosure but a small proportion of the total available number of rats.

All around this slaughterhouse the country was riddled with extensive burrows so that the earth was constantly falling in. In one place the rodents had formed a pathway 500yd long leading to a distant house. It would be impossible to find such conditions anywhere today.

CHAPTER 6

THE TOWN RAT

Probably the most important single aspect in which the town environment differs from the countryside so far as the rat is concerned is its comparative lack of change. The town rat is not faced with those adjustments of food and home as the seasons change necessitating constant movements and resettlement for survival. There are some advantages in this, although these may be outweighed by other aspects of life in the urban environment.

Clear-cut separation of the components of the town is not very easy because properties of various types are found close to each other. It is unnecessary to subdivide too far, anyway, because there is a similarity between city units whatever their nature. A typical city consists of a central sector which contains the chief shopping centre; usually adjacent or intermingled with it is the 'office' part of town with offices existing above ground-floor shops or as separate blocks. Quite separate, as a rule, from this area is the part of town where things are *made* as distinct from the part where they are transferred or sold. Finally, all this is surrounded by suburbia, small units of buildings set among gardens and parks.

Apart from a few scattered parks in cities the ground is

concrete or tarmacadam from one end of the city to the other. Standing on this hard covering are the buildings, separated from one another by alleys and busy thoroughfares. It is scarcely surprising, therefore, to find that where rats exist in city blocks they and their progeny remain in that block spending their lives there and not moving far abroad, provided of course that the block provides the necessary requisites for normal living. As modern towns came into being and as the higgledy-piggledy sections of existing towns are replaced by clear-cut separated units, this trend will probably be accentuated.

Perhaps we should not look at the rat in today's towns without seeing the picture as part of the story of human evolution. In medieval cities, with their narrow streets and buildings linked across the gaps by arches and passages, there must have been a continual intermingling of rats. This may account for the fact that plague spread so rapidly through the cities – presumably the fleas were never short of rats to infect. One must remember, though, that the plague was carried by the black rat, which lived in buildings and was more dependent upon them and their warmth than is the brown rat today. The streets in those days served the additional functions of sewer and council refuse tip and would be one of the rats' chief food sources. Furthermore, rats were treated as a normal part of life and were persecuted less than today so that they could venture out with impunity.

All that is a far cry from today's cities and towns with their clean pavements and streets and their much greater distances across from one side of the road to the other. The brown rat dislikes moving in open spaces: it will get from one side of a room to the other by going round the skirting board. It is hard pressed if it crosses a street and in the back street alleys prefers to scuttle along the base of the wall.

If, then, towns today tend to divide rat populations into a series of units with little exchange between neighbouring blocks, it follows that a block can support rats only if it provides food as well as cover. Food must be present in sufficient quantity to maintain a viable group of rats, and so those blocks that are solely offices or shops selling non-edible goods are unlikely to have much in the way of a rat population. The key points in the centre of the city will be the food shops and food warehouses, hotels and blocks of flats and, of course, restaurants. The rats may in many cases occupy the structure of the buildings but the area round the back where the food is stored and that part where the dustbins are situated will be particularly favoured. Spillage alone round the bins can keep a lot of rats healthy and thriving, and may well be one of the chief focal points of city rat infestations.

Some American research workers made a study of the role of garbage as food for the brown rat.[1] They examined the constituents of garbage bins first of all to find out what edible components there were. They found that in the refuse from a typical residential city block each rat was supplied with 150g (just over 5oz) of edible material per day. Now the minimum energy requirement of rat food is 137 calories per 100g (3oz) of food, and despite the fact that about half of the 150g daily ration fell below the minimum energy requirement of 137 calories per 100g of food, the quantity of food available was clearly in excess of the needs of each rat. It was suggested, therefore, that food distribution, rather than quantity, may have controlled the population being studied. In cage tests in which rats were fed the many foods that were found in garbage, they preferred food on which they gained weight. In addition, it seems that they have a sweet tooth and are averse to curries and other highly spiced foods.

Because of the highly localised nature of town food sources rats probably live in tight little colonies close by. In an experiment in Baltimore, Maryland, some researchers marked rats in the city and later recaptured them.[2] They found that 71 per cent of the recaptured adult males and 77 per cent of the recaptured adult females were caught within 40ft of the original capture point.

Country rats may live in equally small areas but some of them range further afield. In North Germany rats were seen regularly to take eels from the nets at the ebb tide and they came from 2 kilometres away.[3] On these raids they had to swim a 50-metre river and it was estimated that they travelled 6 kilometres each night. In England Drummond[4] has evidence that rats travelled up to two-thirds of a mile each way in a night to feed.

Within the city the brown rat chiefly occupies ground floors and cellars and the black rat, when present, lives in the upper storeys. The black rat is a good climber and has on occasion been seen to cross from one building to another by means of aerial wires and telephone cables. This stratification of the two species when they are found in the same area no doubt helps them to each utilise their own particular niche without undue competition with each other. The black rat is more sensitive to the cold than the brown rat – in North Germany, for example, it prefers a temperature of 38° C (100°F) whereas the brown rat prefers a temperature of 30°C (86°F). Thus we commonly find the black rat living in the central heating ducts of large blocks of flats. In northern Europe it seems always to have lived in buildings and has rarely inhabited the countryside as does the more recently arrived brown rat.

Today in Britain the black rat holds on in ports, where it is no doubt supplemented by immigration from ships, although numbers actually getting ashore must be small, as well as in a few inland towns which were connected to ports by canals. Now that the waterway trade has collapsed those black rats left inland will probably survive because of the increase in centrally heated buildings rather than from replenishment of stocks from outside. In the Port of London both black and brown rats have lived ever since the brown rat arrived.

Rat control has now reduced populations to a very low level, but the rat history of the Port of London is an interesting one. Before 1900 rats and mice had almost free and uninterrupted reign in the dock premises and on the ships and populations were vast. The realisation that much damage and disease was due to rats prompted the authorities to set up an organised system of rat-catching. This began in 1901 and up to the early years of World War II a record was kept of the number of rats destroyed. One must bear in mind that this figure is the number of bodies actually *found* – many more are not found. An encouraging picture of annually diminishing numbers of corpses is not all the story – one should also have an objective assessment of damage to products as well as signs and traces of rats as a verification that the *total* population is being lessened as a result of the efforts, and that one is not just creaming off the unwary rats whose numbers are being replaced.

In Britain the numbers of rats have decreased in all habitats in the past five or six decades, and the Port of London Authority should be commended on their early organisation in this respect. In 1901, the first year of operation, 48,000 rats were caught and this figure excluded an even greater number caught on ships. After a year or two it was clear that the initial onslaught had achieved a state of balance with the rats when the crude methods of control then employed failed to make any further impression on the rat population. By the outbreak of World War I this state of equilibrium was clear and the catch was between 10,000 and 15,000 per year. Just after the war, rat plague was introduced and this, more than anything else, directed attention to rats. By then, official interest had sharpened and the appointment of a few full-time rat-catchers together with organised and improved methods of inspection succeeded in reducing populations to a low level, at any rate judging by rats destroyed. During the 1930s 5,000 or 6,000 rats were killed per annum and it seemed that a state of equilibrium had once again been reached. London, like other ports where black rats abounded, had more black than brown and during World War II there were roughly four black to every one brown rat in the port.

Numbers of rats killed may sound impressive, but unless we know the pre-kill number they may mean little in terms of actual population reduction. American workers have carried out more research on the biology of the rat in urban habitats than have their European counterparts and have worked out techniques for estimating the actual number of rats in an area from the signs of their presence such as holes and droppings. The error is in the region of 10 to 20 per cent, which is quite acceptable for difficult work of this nature. In Baltimore, Maryland, they estimated that the total number of rats in 1947 was 165,000 or one rat per five persons,[5] the distribution of rats being closely correlated with that of buildings in need of major repairs. Repeating the survey in 1949 they found that the total rat population had fallen to around 60,000 or one rat per fifteen inhabitants.[6] This fall they attributed mainly to the fact that there was more money about, which enabled people to carry out home improvements, which incidentally got rid of rats. Other factors that had helped were the improvement in the frequency of collection and in the equipment used for refuse disposal, the rehabilitation of sub-standard dwellings and a more rigid enforcement of the sanitation laws which had cleaned up many blocks.

In New York, incidentally, using the same methods the rat population was estimated in 1959 at 250,000 or one rat for each thirty-six persons.[7] One interesting point was the extreme scarcity of outdoor colonies of rats even in yards with ample food and accumulations of junk. Some colonies did exist, but they were negligible compared with those of the southern cities of the United States.

The rat populations of suburbia are small compared with

those of the central areas of cities, but we must not lose sight of the fact that suburbia probably represents an important link in the chain between the countryside and the city. Any form of rat control in cities, no matter how locally effective, must find some way of preventing restocking from the large, loosely knit populations of the countryside.

The important distinctions between country and town are in the vertical stratification of city rodents and in their lack of mobility owing to the barriers imposed by streets. The vertical distribution of city rats is encouraged by one further factor, which provides a habitat unique in its isolation yet serving as a reservoir for the town rats. Beneath the streets of modernised cities and towns lies an intricate system of linked tunnels carrying sewage from every dwelling place, factory and office block to the sewage-treatment plants. Not all the underground tunnels are the same, for example those from houses are usually 6in diameter earthenware pipes which feed into larger bore pipes. Eventually these lead into 5ft high brick barrels along which men can walk and effect repairs to the larger bore parts of the system. These brick barrels are ovoid in cross section, and rats can find a foothold on the bottom because each row of bricks at the bottom of the ovoid juts out. Frequently, rats run along crab-wise with the fore feet on a row above the hind feet. The earthenware pipes are smooth and not easy for rats to travel in except when the sewage flow is small.

The term sewage implies lavatory waste to most people, but it must be emphasised that a great amount of edible refuse from kitchen sinks, canteens, markets, food shops and restaurants goes down the drains into the sewerage system. This is enough to satisfy the needs of a large number of rats, and for a long time sewer rats were a sharp thorn in the side of the rodent-control authorities. What with the flow of the sewage on the one hand and the smooth-walled tunnels on the other, it might appear to be a difficult environment in which to find food and shelter, yet rats have been able to master this awkward situation most capably. The first requirement is not really too much of a problem, for food gets stranded on the side of the flow and rats can paddle in the shallow flow looking for solid items. Nesting sites are less easy. Rats do not live in the drain itself: instead they find disused pipes, called 'dead eyes', or else live in excavations they make in the region of cracks or poor joints in the sewer. Thus sewers in very good repair present few opportunities for rats to build. When rats burrow out of broken pipes they can, by their excavations in the surrounding soil, cause it to dribble away down the sewer so that a hollow space develops beneath the road which may collapse, thereby causing a great deal of trouble (plate 8).

Conditions in the sewers have certain advantages compared with conditions of surface habitats. The temperature in sewers is generally more even than that above ground, just as in the coal mines it is warmer in the winter and cooler in the summer. There are no predators there, apart from man, so in that respect also it is better than life on the surface. (Perhaps I should qualify that last statement, for although British sewer rats have no predators except man, there are reports that the sewers serving New York have a population of crocodiles and alligators. Apparently this comes about because flat-dwellers find that these reptiles, which are manageable conversation-pieces as babies, soon become too much for the confines of a flat and so flush them down toilets. In the sewers, by eating rats, they apparently flourish.)

It is of course dark all the time in the sewers and rats appear to be less nocturnal than on the surface. Perhaps also because of the lack of climatic extremes, the breeding of sewer rats seems to continue all year round without the seasonal fluctuations found in surface populations. Information on this aspect of sewer rat biology is scarce, however, as there has been no real research into sewer rat breeding.

Adversely, there are several things that make life below ground a hazardous affair and may act to control population size. The distribution of rats depends closely on the presence of food and the ease with which the rats can get at it. Near the summit of a sewer system food scraps are not washed away as quickly as lower down the system and thus a large population can build up there. This part of the system may also be more easily passable by rats because of the low flow. Where food is scarce large populations will not build up.

During World War II it was clearly shown in heavily bombed areas of London how closely the rat population below ground depended on surface conditions. Taking a census of rats at manholes, it was found that there were fewer rats in the areas where a large proportion of houses had been destroyed than in similar but less damaged districts. This high degree of dependence upon food supply from the surface is seen when sewers from markets, canteens and restaurants are compared with those servicing areas with little food waste. The former harbour large rat populations.

Flooding is an ever-present hazard, especially following very heavy rain or thunderstorms. The sewer water level may rise to street level, washing away the rats and drowning most or all of them. Many poisons and unpleasant substances are sent down the sewer and these must present a hazard in addition to methane which results from decay processes in the sewer and which may also kill rats.

The sewer rat population does not exist as one homogeneous whole, strung out through the system, but rather as a series of discrete colonies. Research workers at the Ministry of Agriculture, Fisheries and Food Infestation Control Division at Tolworth deduced that such discrete units exist from a variety of facts: a rat disturbed and chased along a sewer will not go ahead indefinitely but will double back past the chasers' legs rather than go into the territory of an adjacent colony. Also, these workers fed rats on dyed food placed at points in a sewer and by noting the distribution of coloured droppings they could work out the range of movement of a colony. Thirdly, by putting down known weights of bait at intervals along lengths of a main sewer it was possible to map the feeding distribution of the rats and check it by then poisoning and recensusing.[8]

Naturally, control of rats below ground has presented many problems, but by placing poison bait on the benching at manholes it has been possible to control the majority of rats. This is done these days from the surface – bait is placed by means of special dropping tackle so that men do not have to go below. The problem has always been not one of poisoning colonies near manholes but of making contact with those rats living in colonies between manholes. When the 'manhole' colonies are wiped out, territorial expansion on the part of 'inter-manhole' colonies can occur and fill the vacant space. No matter how much time and care is employed, the method of poisoning at access points cannot always give complete control, nor, until some method of carrying poison to those colonies living between manholes is found, will it do so.

Since one might suppose sewer rats are confined to a life below ground, why go to all the trouble of trying to poison them when they cannot affect things on the surface? The fact is that sewer rats are by no means as isolated from the surface as one would like to think and there may be interchange between the above- and below-ground populations. When drains are defective, rats have been known to move daily between sewers and nearby houses and buildings. Where a gap between sewer and surface exists, heavily infested sewers can act as reservoirs of rats which restock surface areas on which much time and money has been spent to exterminate rats. In slum property where houses are condemned and left empty for a long time, the water in the lavatory will dry out and rats may get into the building that way. Again, as buildings are demolished, the drains should be sealed because if they are left open they are an open road by which rats from below can reach the surface. Clearly, the sewer systems are a very important focal point for rats in a city and must be considered in conjunction with any surface control work being undertaken. Wherever sewers exist, it seems that rats will find a home in them and exist as a source of infestation for the cities they serve.

Today the brown rat is the chief sewer dweller, but there have been three recorded cases in Britain of the black rat being found in the sewers: at Wandsworth in London, in Liverpool and in Ramsgate.[9] These three attempts are not the only ones on record, for the black rat has been recorded from the sewers in Peru. However, it seems very unlikely that this species will ever become established in sewers for a variety of reasons.

In general, the harsher the conditions, the more dependent rats must be on the man-made environment for survival. This dependence is seen clearly in Alaska which is one of the most inhospitable areas where rats have made a home. At Nome, the most northerly coastal locality where the brown rat has become established, the physical environment at latitude 64° 30″ is severe and many rats in winter are existing in marginal and submarginal conditions.[10]

Rats are believed to have been introduced there in September 1900, when the great barge *Skookum*, one of the largest on the Pacific coast, was wrecked and its cargo taken off, rats presumably accompanying it. In this locality rats are said to have a high rate of reproduction in the summer so that by the autumn there is a dense population. In view of the high winter mortality, it is reasonable to suppose that a high summer recruitment would be essential.

The winter conditions are harsh: the absolute minimum January temperature is −47°C (−117°F) and days colder than −40°C (−104°F) are common. As a result there are pronounced seasonal movements, rats attempting to find winter refuge in heated buildings with an available food supply.

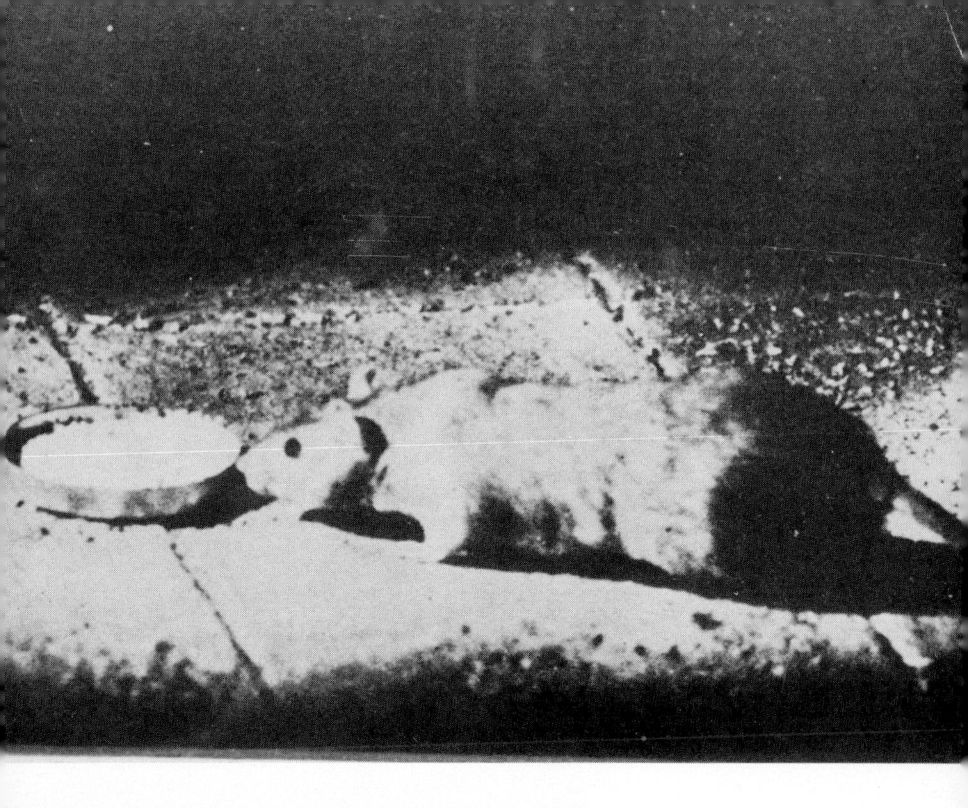

Plate 7 (*above*) Rat approaching a new food source; (*left*) not much left of this cheese!

Plate 8 (*left*) Rat excavations around the fractured pipe caused the collapse of the surface; (*right*) bad urban conditions like these make good rat harbourage sites and impede control

Those that do so survive the winter without much difficulty but those living under less favourable conditions have a tough time. All the rats at Nome are restricted to the modified environment of man in some way or other. At the time of a study made in 1956 the rat population was continuous along the waterfront from a military reservation dump north-west of the city, to the city dump, south-east of the city, a distance of 3 miles. The dumps were centres of high rat population density, another one being in the business district of the city where a burned out bakery which had also housed frozen meat storage in the basement provided food and shelter.

Most rats made for dwellings in winter but the dumps housed many rats despite the weather. On very cold, clear days and nights in winter small clouds of vapour could be seen rising from the openings of the burrows. Although food was scarce the refuse-strewn beach provided cover for rats to move between the population centres.

Cases of frostbite are common among Alaskan rats, the first being seen in late December. By late January the percentage of frostbitten rats had doubled. The damage varied from slightly injured ears and tails at best to severe damage at worst with toes and even whole feet missing. Those with frozen feet appeared to be handicapped when allowed to recover in the laboratory and it is unlikely they would have survived in the wild. As a result of the severe winter the rat population is very low by April.

Despite the harshness of the climate rats have the capacity to make some physiological adjustments to it. Some of the animals caught at Fairbanks during exposure to cold could raise their metabolism more than six times the basal rate during exposure to −40°C (−104°F). This ability of wild rats to supply a high metabolic output in cold may be associated with their very active lives and energetic behaviour. Despite the fact that some rats were exposed to −40°C (−104°F) for almost two hours none suffered frostbite. Since rats at Nome get frostbitten, it is likely that in nature the necessity to get food, especially if competition is keen, forces them to expose themselves to cold often and for long periods.

The vigour of rats in cold warehouses at Nome is impaired and their movements impeded by stiffness or numbness from the prolonged exposure. Rats disturbed in such habitats ran stiffly and crashed headlong into objects. Rat bodies were also found, suggesting that some at least were killed by the cold. Pulmonary infections are severe in some rats and may contribute to the winter mortality. Rats from Nome had fewer parasites than rats from other parts of North America and, contrary to expectation, there was no significant difference in length or thickness of the fur compared with rats from temperate areas. Alaskan rats do not reach the same size as populations in better habitats: the weight range of males at Nome is 28–426g (1–15oz) and females 20–485g (¾–17oz), whereas in Baltimore 8 per cent of 1,432 rats weighed over 500g (17½oz) and in Wisconsin and the Aleutian Islands weights of over 500 and 600g (17½–21¼oz) are not unusual. The author of the Nome study commented that living at low temperatures in marginal conditions either restricted growth or selectively eliminated rats as they aged. The latter looks the more likely because Nome rats become sexually mature at around the same size as rats from Baltimore.

CHAPTER 7

SUGAR CANE, CHEESE AND COAL MINES

There are three main areas in which rats cause damage: growing crops, stored products and man-made installations.

CROPS

Because rat problems in sugar cane are severe I have chosen that crop to treat specifically; also it will enable us to look at the rat problem on a world-wide scale and to get away from any parochial attitudes.

Sugar cane is a grass. Few people in temperate latitudes have seen it growing, but those who have will quickly realise the great difficulty of eradicating any rodent from such a crop. It grows to a height of 12ft in a clump of tough stems each about 1½ to 2in in diameter. It is very leafy and the leaves have finely serrated edges. The cane clumps, or stools as they are called, are grown in rows. Cane in parts of the Caribbean takes approximately twelve months to grow from first shooting to reaping, but in Hawaii it takes rather longer. You can see from plate 3 (*top*) that the ten- to twelve-month crop closely resembles the oft-used term 'impenetrable jungle'.

Pushing through such a mass is a tough business, with the high temperatures, high humidity and teeming insect life. Controlling rats when cane is at this age is a very difficult operation.

In Hawaii the brown rat and the native black rats are the dominant rodents and, quite unusually for a tropical region, the brown rat makes its presence felt in the countryside. If rats are living and breeding within the cane fields, and in Hawaii they may breed all through the year, then a large number will be present by the time the crop is ready for harvesting, and furthermore they will have done a lot of damage by then. If it were possible to be sure of killing off all the rats when the crop is harvested, so that they cannot spread elsewhere, then that would be some consolation but such is not the case: in order to make harvesting quicker (and speed is essential, for once the cane sugar is at peak condition it must be harvested fast or it will begin to deteriorate), the growers get rid of all the cane leaves by setting fire to the field. This sounds drastic but it does no harm to the living cane and it burns off all the rubbish that would get in the way of the cutters. As one might imagine, the rats flee to all points of the compass or take refuge underground: not many perish in the flames, and plenty live to fight and breed another day.

The difficulties for the Hawaiian cane growers are further increased by the fact that their cane fields are situated among forests, near waste lands and adjacent to villages and other agricultural holdings. Many cane fields actually have within them stone outcrops, rock piles and even sections of forest. It is no accident that the number of rats is greater in cane areas that have a high percentage of cane bordering on undisturbed waste land than in large compact fields that have a low percentage of cane bordering on such areas. Many years ago the Hawaiian planters observed that plantations with broken waste land and stone piles within the fields found it more difficult to effect good rat control than did plantations where the ratio of field borders touching waste land to areas of cane was lower. This general pattern is known the world over, not just in respect of sugar cane – any crop prone to rat damage is more severely damaged if it is adjacent to rough, natural areas which can function as a rat reservoir. The rats can flee there as a secondary habitat when the crop is reaped, survive the lean period and return as the crop grows and once again provides a source of food and shelter.

The cane burning hastens the movements of rats from the fields, but in any case they would leave soon after harvesting because the food supply would almost completely disappear overnight. In Hawaii it was noticed that the brown rat stayed in its nests in holes in the ground or in stone heaps for four or five days after the harvesting but that the Hawaiian rat, a form of the black rat, left as soon as the cane was cut.[1] After the harvest the rats migrated to nearby cover or into fields of mature cane nearby which were still untouched. When the next crop was eight or ten months old the rats, which by this time were over-populated elsewhere, came back.

Because of the reservoir in forests and waste lands adjacent to growing cane the first signs of damage in new fields is along the edges. In Hawaii, where there are gulches, the first damage would be at the head of these; later the rats would enter by the steeper sides. As the cane becomes tangled the rats move into the fields permanently, dig holes and build nests. Winter invasions have been a factor in Hawaiian cane damage but no one single factor seems responsible for these. One reason may be the fluctuation in the amount of natural foods. In some seasons drought may reduce the natural food supplies, making it imperative for the rats to shift into the cane. Human activities have not helped, either. Clearing waste land and ditches or cutting strips on the hillsides before planting forest seedlings have all resulted in rats moving to cane for food and shelter.

Just which type of rat is the main cane destroyer depends

on the type of habitat in the proximity of the cane fields. If forest is adjacent then the black rat, a good climber, is prevalent; the brown rat favours human settlements and damage by this species is greatest in fields close to villages, especially when there are pig farms or poultry yards to harbour brown rats.

Rat damage to cane occurs at various times in the life of the crop. When the new shoots first spring from the cane shoots left after harvesting these may be bitten off. New ones will shoot again in a few days but when shoots a month or two old are cut off, even if new ones grow, they will be a long way behind the originals in growth and may never make good sugar. As the stems become woody, rat gnawing, if it does not actually fell the cane, allows fungus and insects to find a way in and add to the losses. Some of the more nimble species of rat can climb the tall stalks and eat the nutritious growing terminal bud: this kills the stem or, if a new growing bud develops, slows down growth and spoils the sugar content. In the most severe cases (plate 3, *bottom*) damage may be so severe that it is not worth harvesting the cane and a field, or fields, may be written off as a complete loss.

With this immensely difficult environment, control is a ceaseless battle. In the early days, desperate planters in the West Indies resorted to dropping poisoned bait packages from the air but so many became lodged in the thick cane that few reached the ground where they could be eaten. A compromise must be reached as a rule: no single measure is usually enough on its own but a combination of tactics may keep the rodent population down to a low level and, while not preventing damage, at any rate might keep it small.

In Hawaii, apart from poisoning, good results were obtained by keeping a clean culture strip, a *cordon sanitaire*, 6 or 8ft wide around the edges of fields bordering on waste areas. Rats were reluctant to cross this open area to reach the cane. Another scheme was to clear shrubs and weeds from the gulches which were a source of infestation and to plant forest trees which shade out the ground weeds that furnished rat food and habourage. Today, herbicides are used to clear unwanted vegetation for this sort of technique although they may themselves bring other problems. In the end it seems that vigilance and prompt action to eradicate breeding centres and population buildup, as well as keeping a sharp look-out for the very earliest signs of rat damage, pay the best dividends.

Natural control methods have some part to play in rat reduction and the mongoose has been introduced into Hawaii.[2] It was imported from Jamaica in 1883 and freed on the islands of Hawaii, Maui and Oahu specifically to fight rats in the cane fields. It was never introduced into Kaui, where the brown rat has always been the dominant rat species. In the late 1920s rat damage on that island was higher than on the others, and this was blamed on the absence of the mongoose.

In the West Indies, which has no indigenous rats but to which the black and brown rats as well as the house mouse were introduced through shipping, the mongoose proved a good ally of the planter.[3] It was introduced in the early 1770s and in ten years spread to Jamaica, Trinidad, Tobago, Grenada, St Vincent, Barbados, St Lucia, Antigua and St Kitts. This sharp predator controlled the rats to a great extent, and it was estimated that in Jamaica alone it saved £45,000 ($110,000) per annum. But there is also a debit side to this story. Having reduced the rats, and hence their own food, the mongoose turned their attentions to killing chickens and very soon smallholders could find little to say in their favour. There has also been an increase in crop insect pests and this may be due to mongoose killing off the vertebrate animals that ate insects. It all goes to show the danger of random introductions that have not been carefully thought out.

By and large, field crops of almost every sort have been damaged and even devastated by rats, except, of course, those that are so awkwardly placed that rats cannot get at them.

Coconuts are a good case: the brown rat is not the most ready climber and cannot get up those high trees. (This does not render them immune, though, because the good climbers of the black rat varieties can reach them and destroy the nuts.)

STORED PRODUCTS

Storing harvested crops does not remove them from rat attentions, as we have seen in the case of cereal stacks. But stacks are really part of the out-of-doors environment and are therefore especially vulnerable to rat attack. Processing crops into the finished, marketable food removes them a little further away from rats especially these days when food is stored in concrete, rat-proofed buildings. In years gone by, when conditions were less hygienic, rats normally occupied food stores and a certain amount of wastage due to their depredations was considered normal.

Damage to stored food can be quite spectacular: the picture of a rat-damaged cheese (plate 7) shows an extreme case, but more often the actual damage to food was much less than that to the containers in which they were packed. Apart from disease hazards, the losses by spillage after cases or bales had been gnawed were usually the greater part of the loss. Very few reliable calculations of actual cash losses to food in store, or of growing crops for that matter, have been made. One of the more objective assessments calculated that in bagged wheat the cash loss was 18 per cent, of which the greater part was loss due to bag damage. With particulate, bagged foods, once a bag or box corner has been gnawed, the food dribbles out and there is much loss in that way. Efficient storage is one answer to this problem for it prevents rats getting to the food easily and makes it easier for warehouse keepers to inspect regularly for rat signs. The pictures in plate 4 show the contrast between food kept in an ideal situation and in the worst possible way, so that rats can do what they like.

MAN-MADE INSTALLATIONS

The damage to growing crops and stored products is all fairly obvious, but a sphere of rat activity often underestimated or not appreciated is the damage to man-made installations. In a coal mine once, the pit was plunged into darkness and power failure by a rat that had had the tenacity to gnaw through an armoured cable, finally short-circuiting the power and giving itself a violent death by electrocution.

Rats can in fact live very successfully and in large numbers many hundreds of feet below ground in the deep, dark recesses of a coal mine. Because of concern at the disease carried by rats and the amount of damage they did to fittings, several years ago the National Coal Board and Safety in Mines Research Establishment conducted a study of the biology of the rat in drift mines.[4]

There are two sorts of mine, the shaft and the drift. In the former the only way to the underground parts is by way of the vertical shaft in which cages or lifts transport materials and men in and out. Entry to the drift mine on the other hand is gained by one or more horizontal or inclined tunnels, the drifts (plate 5, *top*), along which rails carry tubs of materials and coal and along which the miners walk. A drift mine may also have a shaft to the deepest parts but this does not alter the fact that the drift is in direct walking communication with the surface, linking it to all underground parts, no matter how deep or how distant. Rats and mice are not unknown in shaft mines but the problem is a relatively minor one compared with that of drift mines.

A point that is important in determining whether or not a drift mine will be rat-infested is its size. Although many have fewer than ten men, including surface and underground workmen, at the other end of the scale there are large pits with as many as two thousand on the staff.

A small mine might have on the surface a single small building, not much bigger than a fair-sized garage, with only one man on duty, alongside the familiar drift and its rails; down below, working the coal, there could be as few as four or five men. In contrast, in a large mine the surface installations might be called an industrial village in which a great array of processes is going on – workshops and stores, canteens, coal screening and processing plant, shower houses and offices, to say nothing of dumps of derelict machinery that might remain in untouched isolation for months on end.

The surface installations may be adjacent to, or even in, an urban complex, or they might be set in farmland or on the edge of moorland. These differences in location are also relevant to the rat story. Below ground, mines differ in extent and complexity as much as they do on the surface, but basically they consist of a series of tunnels or roadways connecting all the coal faces with the surface and these tunnels are traversed by men and, formerly, by pit ponies.

As mentioned earlier, the two most important needs of rodents when selecting a site in which to settle are food and shelter, ideally as close together as possible. In a corn rick the two needs are fulfilled in the best possible manner, but in a habitat such as a coal mine some compromise must be reached.

I must emphasise that what I have to say in this chapter from now on is the summary of my research, which took place several years ago. Owing to intensive efforts by the Coal Board the situation has for some years been greatly improved. Also, there are no longer any pit ponies, and as we shall see later this has materially helped to reduce rat populations.

After examining a number of mines in the north of England and collecting information from 652 drift mines it became apparent that the percentage of rat-infested mines increased with the size of the mine. Furthermore, the environment around the mine was important in determining the level of infestation. Urban mines had most rat problems, followed by those in farmland, while those in moorland had least rat trouble. This ties in well with what we know of rat distribution generally.

In late summer and early autumn there is a movement of older rats from fields and hedgerows to buildings, and a dispersal of young, weaned rats that are looking for somewhere to settle down after leaving the parental nest. With the onset of cooler weather the assemblage of mine buildings is quickly colonised, and the more buildings there are the better the chance of concealment compared with the one or two small buildings of a small mine where speedy detection is fairly certain. Within the buildings those places with least disturbance and nearest to a food source are usually selected, although on occasions quite noisy places may be frequented. Food sources depend on a canteen around whose dustbins some spillage may occur, perhaps a stable and food stores for the ponies underground, and to some extent on the casual throwing down of scraps from sandwiches. The immigrant population may thus pass the winter but, with the coming of spring, breeding begins and the nucleus expands in size.

Drift mines adjacent to rat-infested buildings show evidence in the dust or mud of considerable rat movements although just why rats go down a drift is not known. Normal exploratory movements may account for it, or the outward movements from over-populated areas may be the answer. Not all rats journey on foot to the underground. Some hitch lifts in tubs of materials or pony fodder by which method they may reach the bottom of the deepest shaft mines.

So far, the activities of men and horses in providing food and shelter have guided the direction of rat settlement and movement; but if this is true on the surface, the interdependence of the three underground is even more marked. There, although shelter is almost unlimited, food is in short supply and is strictly related to assemblages of men and ponies.

Neither of these two is evenly distributed and a study of their distribution is the key to rat colonies underground.

The maze of underground tunnels connects the coal faces with the shaft or drifts, and conveyor belts or tubs carry coal to the outside world and materials inwards. Although miners traverse the whole of the roadways, the majority of them exist in fixed groups at such places as the coal face and transfer and loading points.

In order to determine the precise distribution of rats, small heaps of wheat of known weight were placed at close intervals over the whole mine.[5] Each morning, after the night's feeding, the heaps were taken up and weighed and after a few days the amount eaten daily remained constant. The amount of wheat eaten is roughly proportional to the number of rats, and when it is plotted on a manpower distribution map (fig 6) it is clearly seen that the amount of wheat eaten (and hence the number of rats) is greatest where most men are to be found. Where there are no men other than transients there are no rats. Furthermore, these rat colonies appear to be isolated one from another since over most of the mine there is no movement of rats between colonies and limited movement around individual groups. Rat movement, as human movement, must take place along roadways, and the fine layer of dust everywhere makes an excellent tracking medium for gauging rat activity by means of footprints.

In those mines with underground stables, and hence abundant pony food, the rat populations were extremely high around the stables. (Now that pit ponies are no longer used this source of rats has been eradicated.) Clearly there was a close dependence of the rats on men and horses and the reason appeared to lie in available food. Without men and horses a coal mine would represent an inhospitable environment, almost devoid of food. Even without the ponies, the food taken underground by men for their meals provided a good food source. An average-sized rat can live on approximately

28g (1oz) per day, and as it was not uncommon to see the equivalent of several loaves in the form of sandwich leftovers within the vicinity of a medium-sized group of men, there is little wonder that many rats could be supported. It was not even uncommon for the men to sit down for a meal break to find the rats had been at the food before them. So, although shelter was almost everywhere available, it was the men and horses that determined rat distribution.

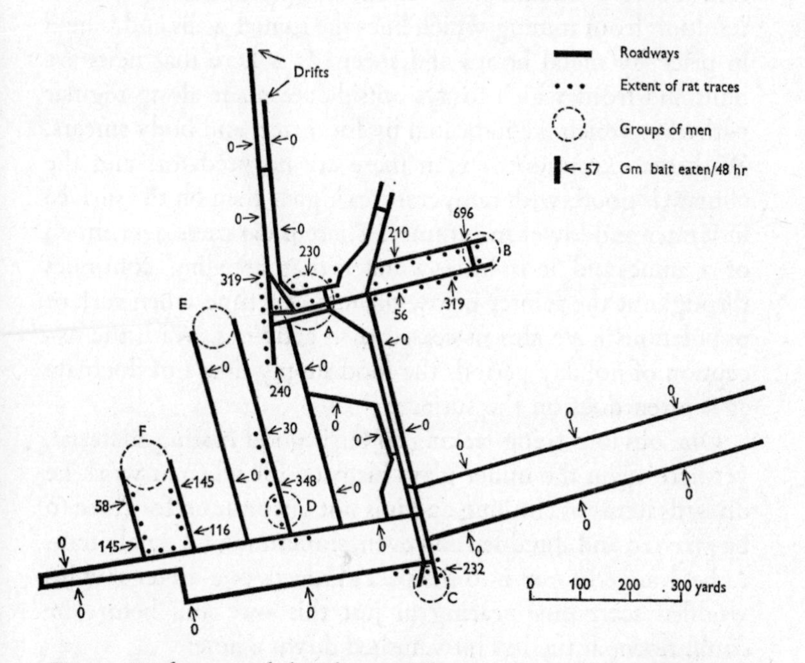

Fig 6 Underground distribution of rats and men. The amount of test bait eaten indicates the size of each rat colony. Note that the rat colonies are closely associated with permanent groups of miners and, although this is not shown in the figure, the size of a rat colony is roughly proportional to the number of men in a group since under normal conditions the rats depend on food thrown down by the men. Hence, the bigger the man group, the more rats it can support

Once a year, though, the rats must have had to tighten their belts, for then the mine closed for a couple of weeks and with the exception of a few maintenance men the pit was deserted. This fact was well known to those responsible for rat control, for it was then that rats would feed avidly on poison bait. One cannot but speculate on the effect on rat wellbeing of a strike persisting for several weeks or even longer.

Given a regular food supply, life in a mine is good for the rats. There is abundant cover in the 'pack', the loose rock resulting from mining which lines the tunnel walls and is held in place by metal hoops and sheets. It is here that nests are built and from which forays outside are made along regular pathways which are indicated by footprints and body smears. With the exception of man there are no predators and the climate is good, with temperatures higher than on the surface in winter and lower in summer. There is darkness over much of a mine and it is little wonder that breeding continues throughout the winter below ground at a time when surface populations have almost ceased such activities. With the exception of holiday periods the food supply does not fluctuate as it often does on the surface.

One obvious thing lacking so far is good nesting material, yet here again the miner plays his part, for when at work he discards items of clothing and it is not uncommon for these to be gnawed and shredded and even, sometimes, for small items to be dragged away into the pack. Indeed, one miner saw his woollen scarf disappearing in just this way and before he could rescue it the last bit vanished down a hole.

Apart from aesthetic objections to such goings on there are strong objections on health grounds since, apart from other organisms, rats harbour in their urine spirochaete bacteria which are the causative organisms of Weil's disease, or leptospiral jaundice, a serious illness with often fatal results which used to be a serious occupational hazard of coal miners. Another important bacterium which may be passed to man is

Salmonella, the one responsible for food poisoning, so that on these two counts alone this close relationship was detrimental to miners.

The coal mines are yet another example of the extreme versatility of rats in utilising any possible environment and adapting their way of life to the situation in hand.

CHAPTER 8

SOME RAT-BORNE DISEASES

Many mammals are well known for carrying disease organisms. Foxes and vampire bats carry the virus that causes rabies, a very serious disease, and many rodents carry the bacterium responsible for bubonic plague. Usually such carriers harbour one species of pathogenic organism, any others they possess being specific to themselves or to closely related species. The rat, however, tends to carry, often within the same individual, several organisms that can cause serious illness or death to men, farm animals or pets. The total list carried by the brown rat is quite long and a detailed account of all the members would take up a lot of space. In any case some are more dangerous and more spectacular than others and as we know more about these I shall concentrate on them.

The one rat-borne disease beside which all others pale into insignificance is bubonic plague, the Black Death of medieval times, which for many centuries ravaged Europe and Asia, causing the death of millions and creating deep, despondent fear in the minds of simple people. The brown rat has never been the main rat species involved in this disease but it can carry it and has done so in many countries and even in the British Isles as recently as 1910.[1] The main rat concerned with plague-carrying on a world-wide basis is the black rat, but the differences between the black and the brown rat in this respect are so slight that what I have to say can apply equally to either.

The history of bubonic plague epidemics goes back to pre-Christian times. Because its symptoms are so well marked it has been carefully documented and we have a wealth of anecdotal and more objective data going back through the centuries. For no other disease is there such a wealth of information; much of it, however, needs treating with caution, because many conditions such as smallpox and influenza were treated as 'plagues' and in the literature have been confused with true plague. However, the broad picture is clear enough.

The causative organism is a bacterium called *Pasteurella pestis*, which is found living in the blood of wild mammals in many remote parts of the world where from time to time plague erupts in epidemic proportions. It is from these wild regions that the disease has spread at various times to most parts of the world.

Provided the plague bacilli remained in the blood of the rodent host all would be well, but certain fleas are capable of transporting them to man. For example, on the brown rat is found a flea species called *Xenopsylla cheopis*. This flea obtains its nourishment by piercing the skin of the rat and feeding on its blood. If plague bacilli are in the rat then some of these will be taken into the stomach of the flea with its meal. In the flea stomach and in part of its gut the bacilli multiply and eventually form a large mass, some of which are regurgitated into the wound when the next meal is taken. In crowded living conditions, in which rats are in very close proximity to men (in medieval times in Europe, or today in Asia and parts of Africa and South America), there is a transference of fleas from rat to man. The flea in due course needs a meal, which it gets by piercing the human skin and taking in blood. During this process flea saliva, which contains an enzyme to stop the blood clotting, passes into the wound, taking with it the bacilli of plague. In the new host they begin to multiply and after a while the symptoms of plague appear. Classical bubonic plague derives its name from the large lumps, or buboes, which appear in the groin and armpit. The swellings are the infected lymph nodes.

Without treatment very many victims die, and in the Black Death that raged across Europe in the fourteenth century it is estimated that one-third of the total population succumbed. Even today, with antibiotics, some patients die but the death roll is small compared with that in the past.

This, then, is the straightforward bubonic plague, which depends entirely on the bite of an infected flea for furtherance of the disease. But there are other variants of plague that are even more severe. Let us assume that ordinary bubonic plague is raging and that Mr X is bitten by an infected flea. In the course of buboes developing, some of the bacilli get past the lymph nodes and find their way into the lungs where they set up infections in the tissue. Mr X will be forced to cough violently, perhaps to sneeze, and in so doing he will spray out large numbers of plague bacilli with the droplets. These, in just the same way that we acquire cold and influenza viruses, will be inhaled by people in the vicinity and will set up plague infections in their lungs. Because of its involvement with the lungs this form of plague is known as *pneumonic*.

The necessity for a flea to transfer the bacilli has now been dispensed with and pneumonic plague can proceed rapidly throughout a population as coughing spreads the disease. It is this form of plague that often occurs in winter in the colder parts of the world such as northern Europe or Asia. Clearly, it is more to be feared than bubonic plague – one person can infect many and the effect snowballs. It was pneumonic plague that broke out in Manchuria in 1922, killing an estimated 10,000 people. Death takes place quicker with pneumonic than with bubonic plague: in the Manchurian epidemic the average expectation of life once plague had made itself evident was 1·8 days. In Europe in the Middle Ages as plague raged on for years, bubonic plague in summer would alternate with pneumonic plague in the winter months.

There is a third variant, which has one thing in its favour – its victims suffer little. The reason is that from start to finish all is over within a matter of hours, many victims, unsuspecting that they have the plague, going to bed at night and never waking. This type is caused by the same organism as the other two but in this case the whole brunt of the infection and the dangerous toxins of the bacteria fall on the blood stream and a fatal septicaemia develops. It was this form that was the cause of so much perplexity among medieval physicians who, in any case, did not understand normal bubonic plague and to whom this last version, occurring as it did in the midst of normal outbreaks, seemed a final devastating stroke from the Almighty.

There have been three major pandemics, affecting vast regions and millions of people. The first one spread throughout the Roman Empire in the reign of Justinian; the second came from Asia in 1330 or thereabouts and raged and simmered throughout Europe until the 1880s; the third began in Central Asia in 1890 and from there spread in shipping along the trade routes of the world. We might in fact consider that today we are in the midst of this pandemic because it is still going on in Africa and America. Plague obtained access to South African ports in 1899[2] and travelled along the lines of communication of the Boer War eventually to reach the *veldt*, where it spread to endemic mammal species in which it still smoulders on, breaking out occasionally and killing its host species. It reached North America by way of San Francisco in 1900.[3] The Rockies were eventually crossed and in 1942 it crossed the Canadian border.

In many parts of the world, including Asia, Africa, North America and Uganda, plague has left its original carriers, the black or brown rats, and found other hosts in the endemic wild species. It is then known as sylvatic plague and represents a continual threat because it ebbs and flows among these wild species and can then get back to rats or those other mammal species that come close to man or even share his accommodation.

From past evidence plague will smoulder on perhaps for several hundred years once it has gained access but will eventually peter out. Thus we are justified in considering this third pandemic to be still in progress, although compared with past episodes its effect is considerably muted.

The important point about transmission of rat-borne disease to man and farm animals lies in the fact that rats may live close to man and his stock. We have already seen how the farm or the coal mine brings the two together and it is not difficult to see how the gap is bridged. One example, investigated in detail, will illustrate just how hazardous this close relationship is to one of the partners. The disease is called leptospirosis, sometimes known as Weil's disease or leptospiral jaundice.

The causative organisms are bacteria of a type called spirochaetes. When they get inside a man, a dog or a cow they act like many other bacteria: they cause fever, produce toxic substances, do a variety of damage and may finally kill. Inside the body of a rat, however, these same bacteria cause no trouble at all – they multiply without harm to the host, who may contain them for most of his life. To biologists this comes as no surprise because in the study of bacteria as well as other organisms such as viruses and parasitic worms like the tapeworm, this is accepted as a part of the biological facts of life.

A species that carries pathogenic bacteria without trouble to itself is called a passive carrier, suffering no ill effects and producing no significant levels of antibody in its blood to kill or neutralise the invaders. A successful parasite, which is what these bacteria are, *must* be able to live in harmony with some species, for if they killed off everything they contacted they would become extinct when they ran out of hosts. This process of adaptation probably stems from a long association between, in this case, the spirochaetes and the rat – they may even have evolved together. Man and cows respond in a very different fashion to the bacilli: they are little adapted to the organisms, possibly because, in evolutionary terms, they have not long been in contact with them.

Fig 7 shows in diagrammatic form how the transmission cycle between carrier and non-carrier may operate. It cannot cover all the possibilities of infection but will be useful to refer to in the following account.

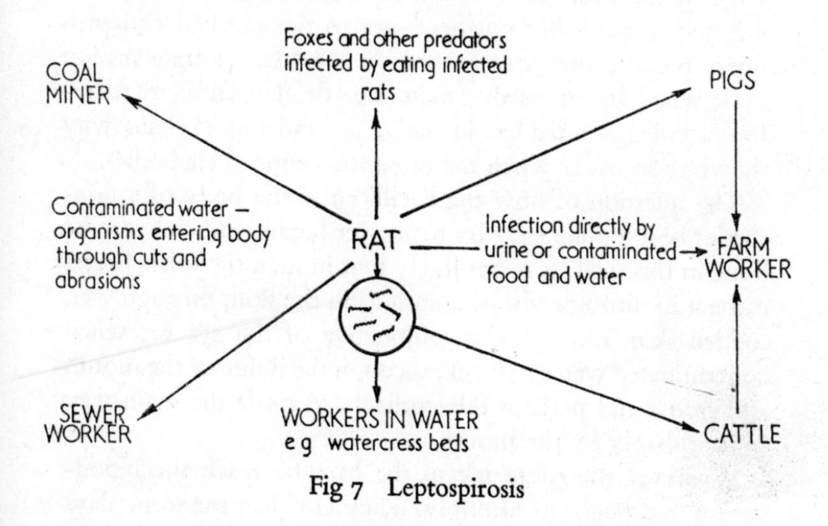

Fig 7 Leptospirosis

We shall begin with the kidney of the carrier rat. A kidney consists of a large number of very fine tubes all twisted and coiled together. The leptospires in the rat live in dense colonies in the kidney tubules and pass down them with the urine to the bladder from where, periodically, they are shed with the urine. This may be in a variety of places – in the cow shed or food store on a farm, anywhere in the farmyard, the farm ditches and hedgerows; on the river bank and the sides of ponds; in cellars and dank basements or on a miner's jacket down the pit. Not all these leptospires will survive, fortunately, because they are killed by the drying out of the urine and in a dry corn stack or in the dry parts of a mine the urine will soon be soaked up or evaporate.

Assuming, though, that a batch of infected urine is released on the wet, muddy bank of a fish pond in rain, the leptospires will be washed into the pond and will survive. If the water is favourable they may multiply, but in any case the crucial stage when most are destroyed has been safely weathered. The next stage in the process, whereby men and farm stock become infected, is probably not necessary to the survival of leptospires because they can survive in a rat-to-rat transmission cycle which has no need of men or cattle. The latter are therefore accidental intruders in the affair, and this explains why they fare so badly when the organisms enter their bodies.

The question of how the bacilli enter the body of a non-carrier host is none too easy to answer because much has to be done in this field. It seems likely that in man the spirochaetes can get in through cuts or abrasions in the skin, through wet, sodden skin, through the conjunctiva of the eye or, when contaminated water is drunk, through the lining of the mouth and throat and perhaps the stomach. In cattle the main way in is probably by the mouth.

Whatever the route taken, the bacteria reach the bloodstream and begin to multiply. They circulate for some days and during this stage the first symptoms make their appearance. These often start quite abruptly in man, with sudden high fever, pains in the thigh muscles, diarrhoea and vomiting. After about a week (in the case of rat-borne leptospirosis, but not necessarily in forms carried by other mammals), jaundice develops owing to liver damage and this is often the point at which the doctor gets a clue as to the nature of the illness. During the second week the leptospires begin to accumulate in the kidney tubules and this is the crucial stage –

if treatment is too late and the patient's own body defences cannot cope with the invaders, death occurs owing to kidney damage which will ultimately lead to kidney failure.

Before the advent of antibiotics the outcome of the disease depended on the patient's natural resistance, and even with modern drugs the result may hang in the balance. Where the patient is severely ill good results have been obtained by putting him on a kidney machine, thus taking the load off the kidneys and giving them a chance to repair the damage. Despite the serious nature of this disease there are no after-effects once recovery has been made, although convalescence may be lengthy.

It should by now be clear that this disease is an occupational hazard in certain jobs of work. If we look at a fairly recent period of time in Britain the figures will show what I mean. In the three years 1947–50, 418 cases of Weil's disease were recorded in England, Wales and Northern Ireland and the breakdown of these by occupations was as follows:[4]

	(*per cent*)
Farm workers	31
Sewer workers	5
Coal miners	4
Food handlers	4
Fish workers	3·5
Workers in water	2·5
Bathers and those accidentally immersed	19
Miscellaneous	31

This shows very clearly the risks to the farmer through rats on the farm in respect of this one disease alone – and there are others not far behind leptospirosis in their potential severity. Among farm workers most cases occur in the autumn and winter months. But all cases except those due to bathing, which occur between June and September with two-thirds in August and September, and in miners, show this pattern when grouped together. The seasonal incidence is obvious in bathing cases but less clear in miners. In an analysis of the occurrence of the disease in Welsh and Scottish miners between 1924 and 1945[5] it looked as though most cases appeared from January onwards, but there were too few recorded cases to make any reasonable assessment of a causal seasonal factor.

Weil's disease has waned among miners in the past twenty-five years for a variety of reasons. Over the years since World War II mechanisation has proceeded swiftly – conveyor belts have taken the place of horses for drawing coal, and better pumping has reduced the number of wet mines and hence conditions essential for the survival of the leptospires. In addition, many drift mines have closed: the risk of infection has always been greater in this type than in shaft mines.

In a survey of the occurrence of the disease in drifts[6] certain points emerged: not surprisingly, each mine was found to be rat-infested at the time of a case occurring; secondly, these mines always had wet underground workings; thirdly, horses or ponies were always underground in all the Welsh pits reporting cases and in the two Scottish mines with the highest number of cases. Finally, with the exception of one surface worker in Scotland, all the cases recorded between 1924 and 1945 were in underground workers. Among the infected miners it was noteworthy that the majority were workers at the coal face. That is where most rats occurred, where the work is toughest and hands and knees get many cuts and abrasions, and where men are in closest contact with the ground where the leptospires are shed. Often it is wet as well.

In the sewers there is an interesting parallel – the incidence of the disease is higher among those men responsible for rebuilding the walls (plate 5, *bottom*) because their work with concrete and bricks causes many hand abrasions, than among the 'flushers' whose work entails less manual labour which could cause damage to the skin. It was interesting that during World War II the number of cases among sewer workers fell sharply in 1940 and remained at a low level throughout the war but increased again in 1946 – during the war there was cessation of extensive repairs and renewals. Thus we can see how human activities in those occupations where rats abound influence the distribution of Weil's disease cases.

Most cases are among men. Of 465 cases in England, Wales and Northern Ireland, 439 (95 per cent) were male.[7] This does not mean that men and women differ in their response to the bacteria, but rather that most workers exposed to rats are men. This high male incidence is usual except among fish cleaners, where most of the workers are women. When the chance of infection is the same there is no difference in the incidence between the sexes. The death rate varies according to the age of the person infected. Under thirty years of age there is a low mortality but between thirty-one and forty-five, and to a slightly lesser extent between forty-six and sixty, the number of deaths is approximately one-third of the number of recoveries.

Human infection is only part of the story. The greatest impact made by rat-borne leptospirosis today is probably on farm animals. It is very difficult to measure exactly how much animal ill-health is due to leptospirosis but serological studies of cattle in Berkshire show a very high incidence of infection under certain circumstances. Cows, pigs and horses probably have a higher resistance than man to leptospirosis and thus we miss cases which in any event are probably often sub-clinical. The course of events in farm animals runs similar to those in man once the leptospires enter the body. In addition, milk production may cease and can be affected permanently. Cows in calf abort and infertility, which may persist, is often a clue to the presence of leptospirosis. The same may happen in pigs and horses.

The influence of the environment on farm animals is an important factor in the incidence of Weil's disease in the herd. For example, half the cows in a herd studied in Berkshire had at some time contracted leptospirosis.[8] The cardinal points in this particular example were, first, a heavy rat infestation around the farm area and, probably equally important, the fact that the farmland was marshy. The two essentials were therefore fulfilled at the same time and the resulting high infection rate was no great surprise.

One part of the farm where animals and rats come into close contact is the pig sty. Pigs grunting and snuffling for food spilled on the damp floor of the sty have a good opportunity to take in the spirochaetes and when they have finished feeding the rats come out and scavenge the bits left over. Indeed, the link between rats and pigs is an interesting one because the two come together in another parasitic sequence, forming part of a cycle that includes man. In this case the parasite is a small round worm of the group called nematodes. These worms live in the muscles of the rat. On occasion pigs eat rats, and when they do the worms go through the gut wall, into the blood stream and eventually settle down in the flesh of the pig where a cyst forms around them. There they remain and the next host will be man, but only if he eats pork that is not properly cooked: then, the cysts dissolve in his stomach and the worms move into muscles anywhere in the body, including the diaphragm or the eye muscles. The cycle back to rats or pigs is completed when infected pork is fed as garbage to pigs and is eaten by rats as well (fig 8). Since rats are often common in pig farms the importance of rat control cannot be too highly stressed.

From time to time we hear of outbreaks of food poisoning, a most unpleasant condition characterised by violent abdominal pains, diarrhoea and vomiting, and which can result in prostration on the part of the sufferer. A group of bacteria under the general name of *Salmonella* is usually responsible for many outbreaks of food poisoning. These organisms live in the rat gut and when the rat defaecates the organisms are carried out with the voided material. If, therefore, infected rats live in food stores or warehouses or in homes, their droppings on or near food can lead to human cases of food poisoning. Or, and this is a more common occurrence, they may be present on the farm, fouling cattle feed with their droppings so that pigs, cows or sheep will be infected and perhaps in turn infect man.

Another intestinal rat parasite, somewhat larger than the last and rather like the harmless amoeba of fresh water, is called *Entamoeba histolytica*. It lives in the rat gut and passes out in an encysted form with a tough capsule around it which serves to protect it in various inclement habitats in which it may find itself. These encysted forms survive outside the body and so long as they are kept moist, can stay alive and infective for several weeks at room temperature. At low temperatures in water or sewage they can remain alive for months. One important point is that they are not killed by standing in cesspools, although treatment in modern sewage works does kill them.

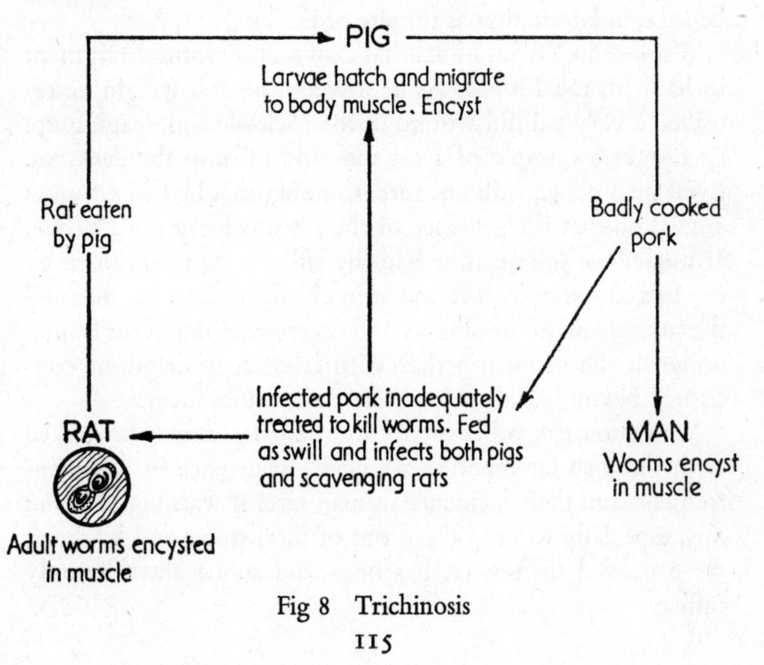

Fig 8 Trichinosis

So far it is apparent that the rat can do severe harm to man without the person concerned ever seeing a rat or even suspecting the presence of rats nearby. Certainly in the cases of food contamination by *Salmonella* or the nematode worm *Trichinella*, the recipient can be hundreds of miles away. In some cases, though, actual bodily contact of rat and man is essential, and in fact the rat must bite the person, before infection can occur. In the salivary glands of some rats live spirochaete bacteria similar to but larger than those causing leptospirosis; these are responsible for a condition or disease called 'rat bite fever'. The bacteria are present in the saliva in the mouth of the rat and when a person is bitten some of the saliva plus spirochaetes enters the wound. The resulting illness is characterised by fever and swellings of the lymph nodes in the area draining the region of the bite. If bitten on the hand the nodes under the armpit will swell or if bitten on the foot or leg then the groin nodes will react. There will in addition be local inflammation at the site of the bite.

A miner in Yorkshire had sat down upon some equipment to have his meal break. As he lowered his full weight he received a very painful wound in the backside and, leaping up, he caught a glimpse of a rat shooting off into the darkness. Even then he was still not sure, thinking that he had sat upon a nail and that the presence of the rat was fortuitous. When, at the end of the shift, he had the afflicted part examined by the first-aid attendant it was seen clearly to be a rat bite and the subsequent bouts of fever and severe swelling of the lymph nodes in the groin together with laboratory diagnosis confirmed beyond all doubt that he had rat bite fever.

Rats also carry other organisms, but they cannot be ranked with those so far mentioned. Their occurrence in rats is infrequent and their incidence in man rare. If warning to avoid rats, especially to keep them out of food stores and kitchens, were needed then what has been said so far should surely suffice.

CHAPTER 9

CONTROLLING RATS

If rat populations are to be effectively controlled, the processes of natural population increase in various habitats and the effects of removing variable numbers of animals from the rat population must be understood.

In order to measure changes in population size or compare population sizes before and after poison treatments we must first of all be able to determine the number of rats in a colony. All sorts of census methods can be used but most of them have some snags or require much experience in their interpretation. One of the most reliable, developed at Oxford in World War II, is a method that enables research workers to obtain accurate field measurements of relative population density.[1] It is known as the surplus baiting method and consists of putting out in containers a known weight of food; this is collected and weighed daily and replaced by a fresh, weighed amount. Thus each day one has a figure for the amount of, say, grain eaten. As we know the average amount of wheat eaten by the average rat in a day we could if we liked use this as a measurement of the *number* of rats eating but there are snags to making this too fine an estimate of the population. It is much better, and none the less useful, to treat it as a level of *relative* population density. As the population increases by breeding so will the amount of wheat eaten increase and the trend can be plotted on a graph. Increase owing to breeding can be distinguished from that resulting from invasion and the technique is a very useful way of studying population changes. Armed with this technique for assessing relative population density, therefore, we can study the way rat populations alter in size in a variety of both natural and experimental conditions.

In any given habitat there will be two main factors governing the ultimate size of the population – the amount of cover and the amount of food available. In some habitats, for example a corn rick, the two requirements are really one and the same and as the population increases and eats the food it will at the same time be reducing the amount of cover. Plate 2 shows what happens during the progressive tenancy of a rick. By the time the rick has been reduced to a sorry heap of straw the rats have run out of cover as well as food and have had to start looking for living quarters elsewhere.

Let us assume that a small number of rats moves into such a habitat and begins to breed. At first population size will increase slowly; it will be essentially a 'young' population in which births are more numerous than deaths. As more young animals come along, mate and produce young then the increase speeds up. In fig 9 the three main phases of population growth are shown. You can see that after the initial slow phase it picks up and over most of the growth curve is what we call 'linear'. (In other words, when population increase, ie actual numbers or relative density, is plotted against time it is more or less a straight line.) As the population density reaches the third stage it slows down and the curve flattens out at the top. This means that the population is stable; deaths are being replaced by births and the total volume of rats, what is called the *biomass*, remains fairly constant. Such a population has reached what American ecologists call the 'carrying capacity of the environment'. This means that the particular habitat has got the maximum number of rats that it can support because there is either no food or no shelter for any more. Unless there is some important change in the habitat the population will remain at this level indefinitely; if the population is reduced but the habitat is unchanged, the number of rats will increase until it again reaches this level.

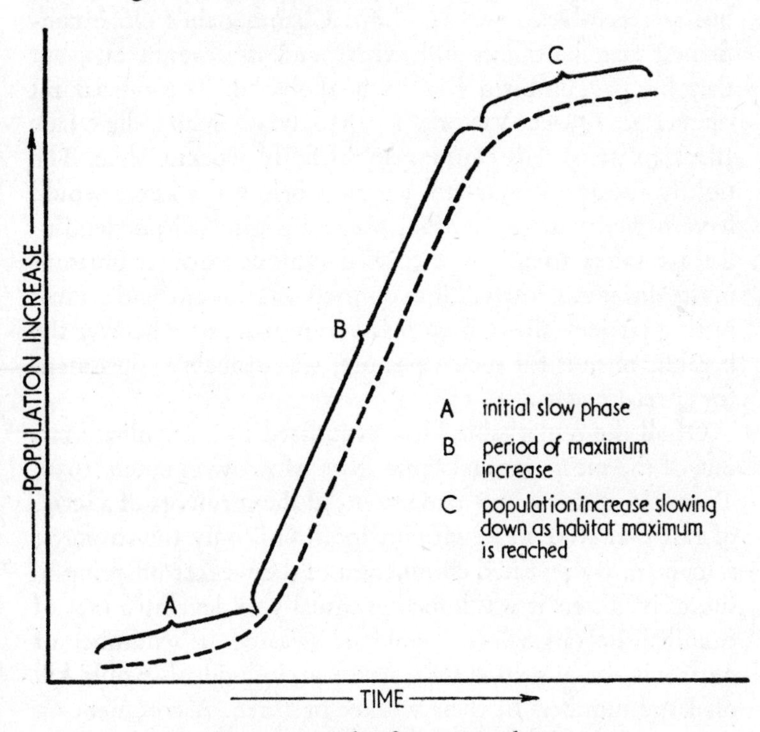

Fig 9 Growth of a rat population

The history of rat control is an interesting and well-documented episode in pest biology. The oldest methods known involved poison and various trapping devices.

The rat-catcher, sometimes a freelance operator, sometimes employed by large organisations and estates, has held a special place in the annals of rat control. Some of them seem to have been colourful characters. In the visual sense of the word, however, probably none was so colourful as the official rat catcher to the Court, an officer of his 'British Majesty' who was bedecked in a special uniform of scarlet, embroidered in yellow worsted, with figures of rodents destroying wheat sheaves. This was in the eighteenth century, but as recently as 1908 the Lord Chamberlain's office confirmed that the officer still existed and still caught rats, but that his gay uniform had been abolished. The official rat catcher to Queen Victoria in 1850 was a man called Jack Black (plate 6). His clothing looks fairly workmanlike. The tightly fitting trousers and leggings below the knees would have been a wise precaution, because frightened rats tend to flee for safety to any dark crevice and open trouser bottoms invite disaster. Clearly, his main tools were a cage and a small ratting terrier – the sash with rats 'rampant', or whatever the heraldic term is for such a posture, was probably adornment for special events.

Of all the remarkable ideas postulated to exterminate rats one of the most unusual came from Moscow in about 1930. There, an attempt was made to breed the survivors of a series of rats that were kept without food until only the strongest remained. By repeated elimination of the weaker offspring in successive litters it was hoped eventually to develop a race of cannibalistic rats which would be released at a number of centres in the hope that these ferocious individuals would kill off large numbers of their weaker brethren. A comment on this novel idea at the time by the United States Department of Agriculture was: '. . . it is to be hoped that the backers of this scheme will not be disappointed by finding that they have developed a super race of poultry killers and baby biters'.

The people of Leeds had their own ideas on rat control (plate 6) but there is no evidence whether this idea is based on experiment, hearsay, or flight of fancy.

In the late 1920s and during the 1930s, poisons had indicated a saving in labour over trapping and their use was on the increase. Had it not been for World War II, however, rat control would have developed less quickly. Soon after war began it was obvious that food storage depots, or buffer depots as they were called, were losing quantities of much needed food for the rationed population because of the depredations of the brown rat and the house mouse. As a result, a group of zoologists was assembled in Oxford at the Bureau of Animal Population to study the three mammal pests in Britain that were responsible for most of the damage to stored products and growing crops (the rabbit makes up the trio). The work of this group made a major contribution to our knowledge of the control of rats and mice and to the war effort.[2]

A great deal of the group's work on the rat was devoted to using a wide range of poisons under a variety of situations and to studying the biology of the rat in order to control populations more effectively. The most valuable finding, which revolutionised poisoning techniques, came about in the pursuit of other aims.

During a poison-baiting experiment the poison was placed in odd-shaped boxes to which entry was obtained from underneath. Rats did not enter these boxes for some days, but eventually they gained confidence and used them freely. (This was perhaps not an entirely new observation for in the Hawaiian cane fields in 1938 R. E. Doty[3] had observed that rats that he was trying to poison were suspicious of a newly discovered food supply and the takes of bait were poor for two or three days.)

Following the reluctance of rats to enter the bait boxes at Oxford, another member of the team noticed that rats showed an initial avoidance of harmless objects placed in their environment. This observation was the beginning of a long series of experiments which revealed many aspects of rat behaviour in relation to environment and showed the relevance of the early observations on reluctance.

The most easily measurable response of rats to environmental changes is the change in the feeding pattern. Rats feed during the hours of darkness, as a rule by making a series of short visits from their place of concealment to the food, taking a small amount and retreating to consume it. In the course of a night they make many visits to the food and thus anything that upsets them will suppress the number of feeding visits and also, of course, the amount of food taken. The effect of disturbances can be measured therefore by weighing the food before and after and recording the amount eaten. Direct observation of the number of feeding visits is obviously another way to record disturbance.

The responses of wild brown rats to various changes in their environment are similar: they behave in a manner that often helps them to avoid destruction but does not enable them to discriminate between potentially harmful objects and harmless things. Because they give the same reaction to all new objects which they encounter in the habitat, the Oxford workers called this response the 'new object reaction'.

Understanding of this behavioural characteristic profoundly modified rat-catching and poisoning. A variety of factors would bring about the new object reaction. For example, in one rat colony the rats were accustomed to approach their food along a certain path. During a normal feeding period they would make a number of forays to the food source and take food: they would also, for various reasons, make some part-way visits. These are easily visible reactions which a hidden observer can record. In the experiment whose results are given in table 2 a piece of wood was placed close by the food in the direct approach path. Before the wood was placed, a control disturbance was created at 8.15. This consisted of making a noise to startle the rats and would contrast with the new object reaction.

TABLE 2

New object reaction: the effect of a piece of wood upon the number of visits to food

Time	*Visits to food*	*Part-way visits*	*Visits elsewhere*
8.15pm	control disturbance		
8.30–8.36	48	13	0
8.40	piece of wood placed		
9.00–9.06	0	6	3
9.06–9.12	0	5	0
9.12–9.18	0	7	7
9.20	piece of wood removed		
9.40–9.46	48	14	0

As you can see, in the period 8.30–8.36 there were plenty of feeding visits. The wood was placed at 8.40 and the effect was quite dramatic. From 9.00 to 9.18 there were no visits to the food itself although there were part-way visits, but not so many in each six-minute period as in the 8.30–8.36 period. What was noticeable was that the rats now made visits elsewhere, presumably in the search for alternative food. At 9.20 the wood was removed and in the period 9.40–9.46 the visits to food achieved their number before it was placed, the number of part-way visits returned to normal and the visits elsewhere dropped to nil.

While the effect of the control disturbance was slight, with the full feeding rate soon being restored, the effect of the wood was well marked. Changes in the habitat that consist of moving familiar objects around a little elicit the same reaction, but it is of shorter duration and less strength than that achieved by placing a new object in the environment.

The sort of disturbances that people expect will upset rats in fact rarely do so. Changes in illumination or loud noises have only a small effect and the animals return and feed normally in a minute or two. It appears that rats are not much disturbed by noises that maintain an accustomed tempo, but noises that occur often but at irregular intervals and with varying intensity and duration do disturb them.

So much, then, for the way the rat reacts to changes in its habitat. Even when he does eventually approach food regarded as a 'new object', he is full of apprehension. In plate 7 his body is crouched low, his belly almost on the floor; his approach is very slow, almost stealthy; his head stretched out, and his posture indicates that he is waiting for the worst and ready to turn tail at the slightest provocation. If nothing untoward happens he will eventually take food but feeding will be sporadic at first, only very small amounts of food being taken.

With the poisons that were in common use at that time it was absolutely essential that the rats should feed voraciously in order to take in a lethal dose of poison. Any rat toying with its food therefore might begin to feel ill and hence give up feeding before getting a lethal dose. Since, as we now know, the memory of a nasty poison experience can be retained by the rat and associated with the base with which the poison was mixed, it would be of little use giving either that bait or that poison again.

The practical steps towards better rat-killing that an understanding of new object reaction brought about were really twofold. First of all it was now clear that putting a heap of poison bait down in the places used by rats would in itself bring about avoidance, since the bait was a new object so far as the rats were concerned. Also, whereas it used to be generally thought that sweeping up and moving objects before laying poison was in some vague way desirable, it was now obvious hat the best results would be achieved only by leaving things ust as they were.

As a result of these findings the technique of pre-baiting came into being. In a place where rats were to be poisoned someone would first go round and put down heaps of bait without poison. The next day if much of the bait had been taken he would top them up, taking care not to put the food in a different place, but only on the existing pile. After a period which varied according to the abundance of alternative food sources, feeding would begin, slowly at first, and then increasing. As confidence grew, each night the rats would begin feeding earlier and more voraciously. When this state of affairs was reached the operator would then mix poison with the bait. If the pre-baiting had been carried out well there would be a successful kill with few survivors.

Fig 10 shows how pre-baiting works. The yardstick of activity is the number of visits, which is shown on the left, vertical, axis. The horizontal axis gives the time from dusk at 5.30 to after midnight. These rats have had a new source of food placed in their environment. On Day 1 feeding does not begin until 8pm and continues at a low level until after midnight. With each succeeding night, feeding begins earlier and increases in intensity until Day 4, when it begins just after dusk. Feeding is uninhibited and concentrated so that it is nearly all completed by 8pm, the final feed taking place by 10pm. Poison placed in the bait on the fifth evening would have achieved an effective kill.

Before moving on to recent innovations I would like to mention the relevance of rat population growth to rat control methods. In the graph (fig 9) population increase is slowest at the bottom and the top parts of the curve (A and C sections respectively) but over the whole part of B it is maximal. If a poison programme succeeds in reducing the rat population only moderately, ie between 50 and 90 per cent of the total, the rats show early signs of recovery and increase at a rate of about 4 per cent (between 2 and 6 per cent) of the capacity level each month. As they approach the maximum for the habitat the increase slows down to a rate of about 2 per cent or less per month. On the other hand, if a population is reduced by more than 90 per cent it will recover at a slower rate of between 1 and 3 per cent per month until it reaches a

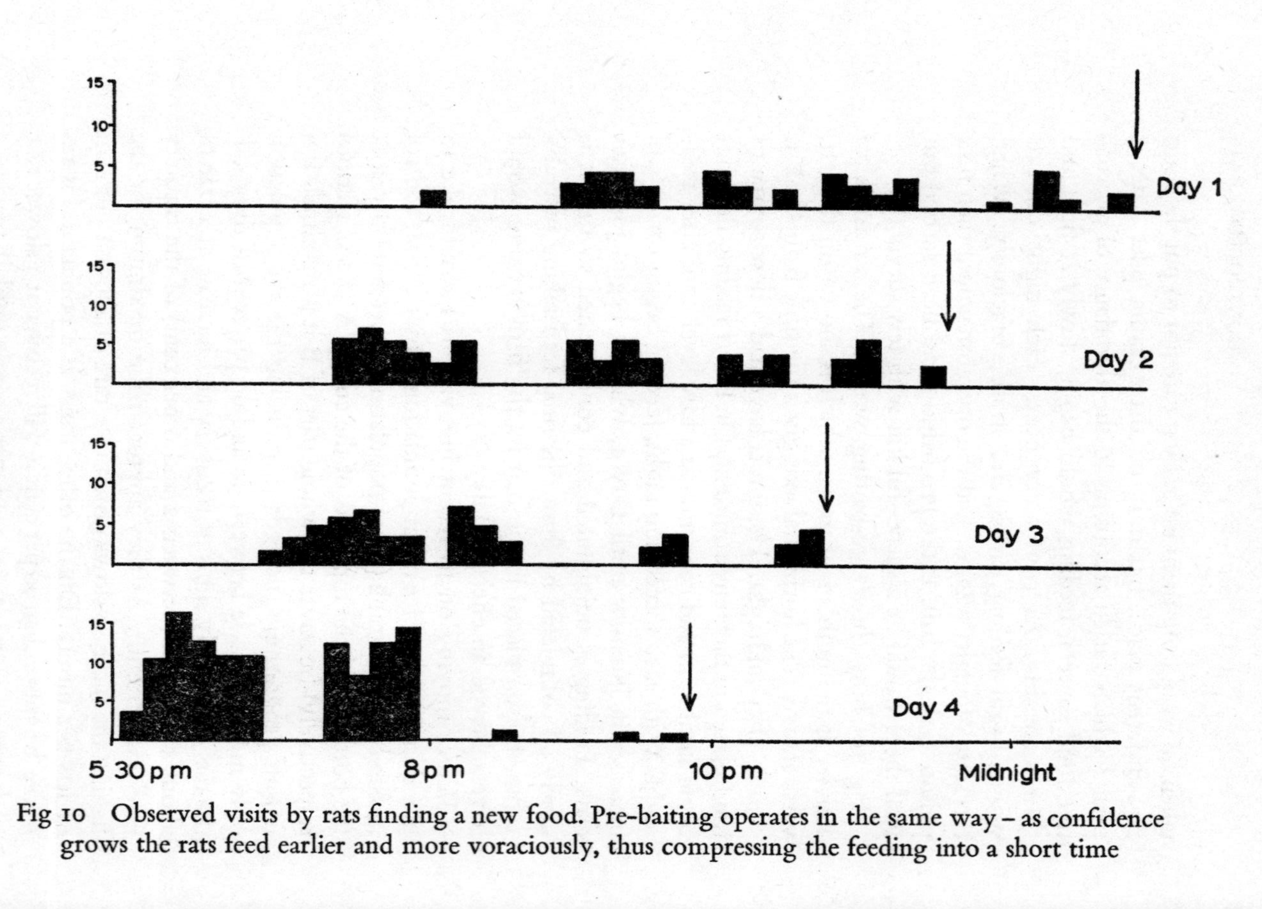

Fig 10 Observed visits by rats finding a new food. Pre-baiting operates in the same way – as confidence grows the rats feed earlier and more voraciously, thus compressing the feeding into a short time

population level which is about 10 per cent of the original, and it then speeds up.

It is thus of the greatest importance in control work to reduce the population by 90 per cent or more at the first poison treatment and then, with a change of poison and bait, to attempt to remove the remainder. A poison treatment that reduces the rat numbers by only 50 per cent is, by and large, a waste of time.

In practice, effective control with acute poisons was rarely, if ever, achieved on one occasion, and lasting reductions of village rat populations were achieved only by comprehensive double or triple strikes. Most rat destruction, unless rigidly controlled to make sure that the job was carried out in a painstaking manner, reduced the rat population only to a point on the growth curve at which the rate of increase was at or near the maximum. This was pointed out by Barnett and his co-workers in 1951[4] when describing a single strike which was carried out in the autumn of 1948.

The battle against rats was waged throughout the latter part of the war using the new techniques, but it was not until 1949 that the next important step forward in rodent control methods was taken. During the 1930s a widespread disease among cattle in the United States was studied. It was called 'sweet clover disease' and the chief effect was severe, often fatal, haemorrhaging. The cause of this was a substance called dicoumarol which was produced in spoilt clover hay. In 1942 a substance marketed under the name of Warfarin was first synthesised as a derivative of dicoumarol and by 1949 it was realised that this substance had excellent value as a rat-killer. It changed rodent control yet again and in the years following its discovery more than one hundred related compounds were synthesised.

These dicoumarol substances are chemically rather complex, but their effect is to prevent the normal clotting of blood. When tissues are damaged either on the outside or the inside of the body, blood usually flows for a time and then clots. This clot prevents excessive blood loss and is part of the normal way the body has of repairing itself. (So as not to go into the fine detail, which would take too long, I shall have to use terms that will not be self-explanatory. Those who are sufficiently interested will at least have enough information to go further should they wish.)

There is in the blood plasma a substance called pro-thrombin. Damaged tissues release a substance which, in the presence of calcium chloride (which is present in the blood), converts the pro-thrombin to the enzyme thrombin. This thrombin, in haemorrhages, coagulates a blood protein called fibrinogen into a strong network of fibrin strands (the blood clot), which prevents further bleeding.

Without pro-thrombin, clotting cannot take place. The role of coumarin derivatives is to prevent its formation. As a result, the small internal haemorrhages that are constantly taking place do not clot and eventually so much blood drains away that the animal dies.

After the action of dicoumarol was discovered it was soon seen to be a very useful poison, with certain singular advantages over the acute poisons it was soon to supersede. First of all, instead of being dependent upon one quick, large feed for its effect, it could be nibbled away at for seven to ten days; indeed, since the effect was cumulative, daily feeding for a week was much more effective than one big dose. Secondly, and this was very important, as the rats began to feel unwell they did not associate their indisposition with the food they were taking, so that there was no question of bait shyness. They had in any case eaten a lethal dose before feeling unwell.

Thirdly, the field use of this type of poison could be carried out with a reduction in manpower. With acute poisons the pre-baiting and final poisoning necessitated a person topping up bait points each day for four, five or six days. With the new poison, provided the operator could effectively gauge the population size, it was necessary only to top up the poison bait points each week. The operator had only to make sure that there was always a surplus of poisoned food so that continual daily feeding could take place. New object reaction was now unimportant in baiting because, although the heaps of bait might induce the response, the rats could be left to overcome their shyness and the taking of initial small doses would not matter.

Poisons using these derivatives were grouped under the term of anti-coagulants because of their action in preventing coagulation of the blood, and they were marketed with the trade name of Warfarin. Their use was soon widespread and they found favour in a variety of habitats if for no other reason than that they were so much safer to other animals. One of the dangers of using acute poisons such as arsenic had always been that of poisoning other creatures either directly, by their eating the poison, or secondarily, by their eating poisoned rats. Because the concentration of Wafarin in bait was so low, usually 0·025 per cent, a large animal such as a cow would have to find and eat a very large number of baits to get enough to do any harm and the chances of that happening are remote. For the same reason, secondary poisoning was almost non-existent.

Warfarin-type poisons were used with great success against rats and other rodents all over the world – in cane fields in Hawaii they used permanent bait stations and in British coal mines the personnel engaged in rat control could now cover more ground than before. In the confines of a dark mine, where men were liable to touch poison bait accidentally, it was safer to use anti-coagulants than acute poisons.

Although it was known that individual rats could acquire resistance to some acute poisons through repeated small doses of a poison eaten during several poison treatments, this type of resistance was of the acquired type and was not a great problem. Indeed, so rosy did the rodent control scene look that in 1953, soon after anti-coagulants were in general use, the United States Public Health Service said: '. . . the possibilities of rodent control appear to be more encouraging than ever before, for unlike insects, rats do not seem to develop resistance to poisons'.

Alas, this optimism was relatively short-lived, and the current problem in rodent control today is the phenomenon called Warfarin resistance, which differs from the acquired, physiological tolerance type of resistance mentioned above in that it is inherited from one generation to the next.

The first such case of resistance recorded[5] was in a population of brown rats near Glasgow, which refused to succumb to anti-coagulants. Two years later resistance in brown rats in the counties of Montgomeryshire and Shropshire on the border of Wales and England was reported,[6] and in 1962 brown rats in Jutland, Denmark, showed resistance.[7] By 1969 the area of resistance in Denmark had increased to 1,000 square kilometres (386 square miles). In the old parishes where resistance was first seen, 48 per cent of the rats were resistant compared with only 6 per cent in parishes where resistance was newly reported.

In all three areas the circumstances were similar. There were reports of failure to control rats on a single or a few farms where large quantities of Warfarin had been used. A variety of reasons for the failures were put forward, the principal ones being that the control technique was faulty or that the natural rat food contained a high content of vitamin K (this enables them to tolerate Warfarin). When rats from these resistant areas were taken into the laboratory and fed Warfarin, the picture of real resistance eventually emerged. Normal rats show a 98 per cent mortality when fed for four days on bait containing 0·005 per cent Warfarin, whereas resistant rats survive long periods on this level of poison feeding – in Denmark resistant rats have been fed for two months exclusively on a bait with 0·05 per cent Warfarin without any obvious ill effects. This resistance is inherited, and aspects of the genetic side of the story are still being studied.

It seems that when anti-coagulants are removed from an area in which resistance exists, and acute poisons are then used, the incidence of resistance will decrease. In Shropshire the zone of resistance has expanded at the rate of about 4·6 kilometres (nearly 3 miles) radially per year, however, and despite intensive control measures both there and in Scotland, it has so far been impossible to eliminate the resistant population by acute poisons.

While poisons remain the most effective way of killing rats in large numbers other methods may still be used. Trapping is still useful on occasion – one would not use it for large infestations, but it has a place in the elimination of smaller numbers and to get rid of the survivors of previous poison campaigns. The most useful type of trap is the break-back, which is set at right angles to rat tracks. As a new object will elicit the avoiding reaction, the traps must be put in position for a few days without being set in order to condition the rats to run over them.

Other methods are gassing and fumigation. The most usual gassing technique is to use a powder that produces hydrogen cyanide when it comes into contact with anything damp such as soil or air. It goes without saying that hydrogen cyanide will kill the operator as well as the rats and very great care must be taken when using it. The usual method of operation is to put the powder down the rat holes either with a long-handled spoon or with a hand-operated pump. The holes are then blocked with turf or soil and any that have been re-opened next day are treated again.

Fumigation is carried out only by servicing companies which are specially equipped and trained for the work. Most fumigation is carried out on rat-infested ships and occasionally in buildings where rats are living in the fabric of the building and are difficult to deal with using poison bait. The most commonly used fumigants are hydrogen cyanide and methyl bromide but carbon dioxide and ethylene oxide have at times been used.

It is known that rats produce sounds that are called ultrasonic – that is, they are so high pitched that human beings cannot hear them. Rats use them to communicate with one another, especially the very young, and perhaps to locate objects in the way that bats do, that is by echo-location. Some experimentation has gone on in recent years to see whether machines that produce ultrasonic sounds have any part to play in rodent control. At the moment it seems unlikely, but they could have a role when we have learned more about the role of ultrasonic sounds in the life of the rat.

Currently, the thinking in rat control is centred around large-scale planning towards a strategy of area control rather than the piecemeal treatment of individual properties. This may be easier for towns than for the countryside and it will in any case be difficult to keep towns free when there is the reservoir outside the town boundary: indeed it may be impossible to achieve rat-free towns until the surrounding rural areas can be made rat-free simultaneously.

There is very good evidence that the most effective control measure that can be practised against the rat is to deny it access to food sources other than what it would normally find in the wild and to prevent it entering human habitations, barns and stores. Some of the most striking improvements in reducing rat numbers have been brought about by simply tidying up the environment. In Baltimore, Maryland,[8] research workers carried out a detailed tidying up operation in a group of city blocks, removing all heaps of rubbish where rats could nest and cleaning up food scraps around dustbins. In nearby blocks with identical features and the same rat problems things were left as they were, to act as a control to the experiment. The rat populations were assessed using a standard technique. As a result of cleaning up the environment the rat populations of two city blocks actually declined by 50 per cent and 75 per cent. The work indicated that the rats did not merely move to an adjacent block but did in fact die.

As long ago as 1920 the proofing of buildings, food protection and garbage removal were advocated as essentials in preventing infestations, yet we have still a long way to go in this respect. Plate 8, *right*, shows the type of situation that can provide living accommodation for a large number of rats. Unless these sorts of situation can be avoided, much of the hard-earned expertise in using poisons will have been wasted.

In rural areas, food scattered on the ground where small numbers of poultry are kept acts as a valuable focal point for local rats which will move in and colonise the area round about. It was shown clearly that in Norwich during World War II even the backyard keeping of poultry or rabbits caused a marked increase in the rat populations.

These problems, together with the proofing of buildings, must be tackled first in any scheme of rat control.[9] If they are conscientiously carried out the back of the problem will have been broken, but the co-operation of all members of the public is essential and this can be attained only by making people aware of the dangers involved in harbouring rats.

CHAPTER 10

PROSPECT

That the rat is a pest, either within the terms of popular usage or the more carefully defined terms of the scientist, there can be little doubt. In general this term is used rather loosely to describe any animal that causes damage to both animals and crops. One dictionary definition of a pest is 'an organism inimical to man'.

Today's careful costing of control operations, however, has refined the term with the addition of the prefix 'economic' or 'controllable'. This is a valid point, for the crop grower is chiefly concerned with how much an animal will affect his crops. From this approach has emerged the term 'economic threshold', which is the point at which a pest can be controlled at a cost less than the expected market value of the potential increase in yield. If economic control measures are not available then the pest comes more within the category of natural calamities such as bad weather and it is perhaps best to leave it so rather than spend a lot of money on uneconomic control measures.

The exact point at which we regard an animal as qualifying for pest status has not received universal agreement but many would consider it to be when there are sufficient animals to cause a 5 per cent loss in yield.

Terms like *economic threshold* and *pest status* imply a degree of accuracy in damage assessment as well as the capability to measure accurately the population size of the pest. These terms have arisen mainly in entomological work on pests because most of the research on crop pests has been on insects, the major crop pests. Techniques of accurate damage assessment, animal number estimation and the forecasting of damage periods have been well developed on a scale unmatched by workers on vertebrate pests. Many insect pests are highly specific in the crops they live on and eat and are highly regular and hence predictable in such things as emergence times. They are therefore more amenable to study than pests like the rat which are omnivorous, well scattered and non-seasonal. There are very few reliable estimates of the actual cost of damage by the brown rat, for example.

The rat is, along with the house mouse, perhaps the most cosmopolitan of all pests, vertebrate or invertebrate, and the great adaptability that coverage of such a range of habitats reveals may in the long run make it a greater pest than any other animal. We must remember that populations of pest animals develop on two time scales, a long-term or evolutionary scale in which a species changes by mutations, and a short-term one reflecting the changes that take place on a seasonal basis. In an animal such as the rat the pest status will increase considerably on farmland during the breeding season, and with the passage of summer to reach a maximum by autumn.

An animal may achieve pest status quite suddenly owing to agricultural techniques. For example, an insect species living in pasture may exist in small numbers simply because only a small proportion of the plants is actually suitable for food, yet will thrive and produce large numbers in a field where a single crop is grown that is ideal for its food. Thus the system of monoculture is frequently responsible for producing pests where few existed before. Likewise, food storage provides an environment that is ideal for stepping up the production of a species that existed at lower levels out in the field. Similarly rat numbers are low in areas of woodland and moorland, become higher in mixed farmland and may be very high in cereal-growing areas, especially when corn ricks and modern grain stores enable breeding to take place all the year round.

The brown rat seems adequately equipped, therefore, to survive almost anywhere. It can live at low population densities with the population just 'ticking over' so to speak, and when conditions improve it can make the most of them.

With this background of capability before us, let us attempt to look into the future and speculate on how the changing pattern of human societies will affect the rat. We already have some knowledge of how today's trends in some quarters are aiding the house mouse. In large cities the great expansion of huge modern blocks of flats and offices, with their networks of central heating pipes running through interconnecting ducts to all parts of the building, has provided the mouse populations with an environment ideal for their spread. Since nearly all parts of the system are more or less inaccessible to man the mice can live unmolested, provided they can obtain food and water, with the result that the mouse is the biggest rodent problem in urban conditions in Britain today.

It is unlikely that the brown rat will become a serious pest in these conditions, because manoeuvring in the restricted and often vertical confines of ducted piping is more suited to the climbing abilities of the mouse or the black rat. Dilapidated urban property, however, has always been a rich environment for the brown rat and so it thrives in the squalid tenements of the world's great cities. There seems but little hope that slums will be eradicated anywhere for a long time to come, and in fact they may be increasing. We may, therefore, expect the rat to continue to be a serious problem so long as these conditions exist.

On a world-wide scale the major biological problem is that of producing enough food, and the agriculturalists are tackling this in two ways. The first is to make the existing cultivated ground produce higher yields, either by improving crop plant strains or by shortening the growing season; the second is to cultivate marginal or unused land. Either way pests stand to gain, and the opening up of new areas might give the rat another lease of life in terms of its geographical distribution or even its reproductive physiology.

Stock rearing is today being carried on in a more intensive fashion especially in those countries where farming has become more mechanised. The housing of poultry, pigs and cattle in ever-increasing numbers has already assisted in concentrating rats and, worse, enabling the readier transmission of some of their parasites to the animal whose quarters they so readily share. Indeed, the whole emphasis in western countries is towards ever larger, more tightly knit assemblages of people and farm animals. Even transport systems are moving towards the conveying of ever larger numbers.

Today one of the most challenging problems faced by medical science is the development of resistant strains of bacteria and other micro-organisms. Many of these have arisen through the use, or misuse, of antibiotics; and unfortunately, because bacteria multiply every twenty minutes in favourable circumstances, the resistant strains arise much more quickly than the microbiologist can devise new, effective antibiotics against them.

But resistance to antibiotics is only one aspect of the ways in which bacteria may change their biology: there are other subtle ways they may alter, often very swiftly. We know that a micro-organism may exist in a mammalian host which is so well adjusted to it that it is inconvenienced by it only very little or not at all. Yet this same organism, when transferred to a closely related species, may change to a virulent form having serious effects on the new host and on other, unrelated, hosts.

The history of the spread of yellow fever is one that portrays such a shift.

Mutations in micro-organisms may be brought about by many factors, for example radiation from radioactive materials, chemicals in the environment and the effects of biological systems; such changes are an inevitable part of the process of evolution. We know that plague has had periods of resurgence, and despite the fact that it is at the moment reasonably quiescent, mainly because health authorities all over the world keep a careful watch on both the rodents and the vectors, there is absolutely no reason why a new form of extraordinary virulence owing to one of the above factors should not appear somewhere in the world.

Given that this could happen, imagine how modern travel could assist the spread. Take for example a jumbo jet capable of carrying 200 passengers loading up in Singapore. Suppose that one of the passengers, an oil technician, only just made it because the helicopter flying him direct from the jungle rig where he had been living was late. The infected flea he had unknowingly picked up in his tent bit him earlier that morning, but in the rush and heat the irritation was slight and after a tough spell in a jungle camp one more bite makes little difference. During the long haul from Singapore to Delhi a fever begins and he feels he has a cold coming on, a conviction enhanced by his sneezing now and again. In the confined space of a jet cabin a few sneezes contact a lot of people, and within a few hours half the passengers have received droplet-borne infection from this one man.

The modern jet touches down for refuelling, leaves a few passengers and takes on some more. At each touchdown point the infected persons catch other planes going in all directions. Within forty-eight hours those original 200 will have scattered to many countries before the oil man, who became critically ill in Athens, has died.

In the face of pneumonic plague, always the most dangerous for whole populations, a world-wide pandemic could swiftly follow. As the final straw, if no antibiotics affect the new strain and laboratory research has to develop new drugs, which might take many months, then the death toll could be so vast that all services would break down and modern civilisation be totally disrupted. This is a black picture but quite a possible one, and many biologists share these fears not only about plague but about other diseases as well.

In the event of such a pandemic and population decimation there can be little doubt that those members of the animal world that would benefit most would be the ones that were least specialised in diet and way of life. The rat would clearly be in a very strong position, and it takes little imagination to visualise how its numbers would increase once the chief predator had been reduced.

Whatever the future holds, it is difficult to visualise an environment that is totally inhospitable and untenable for an animal as tenacious and adaptable as the rat. It must therefore continue to rank as an important world-wide pest and as a biologically important animal for a long time to come.

NOTES AND REFERENCES

Chapter 1

1 Watson, J. S. 'The Melanic Form of *Rattus norvegicus* in London', *Nature* (London), 154 (1944), 334–5.
2 Smith, W. W. 'Melanistic *Rattus norvegicus* in Southwestern Georgia', *Journal of Mammalogy*, 39 (1958), 304–6.
3 Figala, J. 'Yellow Mutant in the Brown Rat, *Rattus norvegicus* (Berk.)', *Acta Societatis Zoological Bohemoslovenicae*, 28 (1963), 209–10.

Chapter 2

1 Fitter, R. S. R. *The Ark in Our Midst* (London: Collins, 1959).
2 Barrett-Hamilton, G. E. H. *A History of British Mammals*, Vol 2 (London: Gurney & Jackson, 1913).
3 Shrewsbury, J. F. D. *A History of Bubonic Plague in the British Isles* (Cambridge: CUP, 1970).
4 Ziegler, P. *The Black Death* (London: Collins, 1969).
5 Chaucer, G., *The Canterbury Tales* (Penguin Books, 1972).
6 Pallas, P. S. in Steiniger, F. *Rattenbiologie und Rattenbekämpfung* (1831; repr Stuttgart: Ferdinand Enke Verlag, 1952).

7 Pennant, T. *The British Zoology* (London, 1761–6).

8 Smith, R. *The Universal Directory for Taking Alive and Destroying Rats and All Other Kinds of Four-footed and Winged Vermin in a Method Hitherto Unattempted: Calculated for the Use of the Gentleman, the Farmer and the Warrener* (London: printed for the author, 1768).

9 Silver, J. 'The Introduction and Spread of House Rats in the United States', *Journal of Mammalogy*, 8 (1927), 59–60.

10 Tryon, C. A. 'Entrance and Migration of the Norway Rat into Montana', *Journal of Mammalogy*, 28 (1947), 188–9.

11 Ecke, D. H. 'An Invasion of Norway Rats in Southwest Georgia', *Journal of Mammalogy*, 35 (1954), 521–5.

12 Brooks, A. 'The Brown Rat, *Rattus norvegicus*, in British Columbia', *Canadian Field-Naturalist*, 61 (1947), 68.

13 Brown, J. H. 'Alberta: The Only Rat-free Province in Canada', *Canadian Journal of Public Health*, 39 (1948), 367–74.

14 Rowe, F. P. 'Notes on Rats in the Solomon and Gilbert Islands', *Journal of Mammalogy*, 48 (1967), 649–50.

Chapter 3

1 Chitty, D. and Southern, H. N. (eds). *Control of Rats and Mice*, 3 vols (Oxford: Clarendon Press, 1954).

2 Morgan, C. T. 'The Hoarding Instinct', *Psychological Review*, 54 (1947), 335–41.

3 Bindra, D. 'What Makes a Rat Hoard?', *Journal of Comparative and Physiological Psychology*, 41 (1948), 397–402.

4 Bindra, D. 'Motivation for Hoarding', *Journal of Comparative and Physiological Psychology*, 41 (1948), 211–18.

5 Barnett, S. A. 'The Feeding of Rodents', in Harris, K. L. (ed). 'Indian Rodent Symposium' (New Delhi: USAID, 1970), 113–23.

6 Karli, P. 'Norway Rat Killing White Mice', *Behaviour*, 10 (1956), 81–103.
7 Steiniger, F. 'Soziologie der Wanderratte', *Zeitschrift für Tierpsychologie*, 7 (1950), 356–79.
8 Harting, J. E. *Vermin of the Farm* (London: Spottiswoode, 1892).
9 Eibl-Eibesfeldt, I. 'The Behaviour of Rodents', in Kukenthal, W. *Handbuch der Zoologie*, 8 (1958), 1–88.
10 Steiniger, F. 'Revier und Aktionsraum bei der Wanderratte', *Zeitschrift für hygienische Zoologie und Schädlingsbekämpfung*, 39 (1951), 33–50.
11 Calhoun, J. B. 'Influence of Space and Time on the Social Behaviour of the Rat', *Anatomical Record*, 105 (1949), 28.
12 Hurrell, H. G. 'Ingenuity of Rats', *Western Morning News* (12 May 1970).

Chapter 4

1 Beach, F. A. 'Copulatory Behaviour of Male Rats Raised in Isolation', *Journal of Genetic Psychology*, 60 (1942), 121–36.
2 Jenkins, T. N. 'The Effect of Segregation on the Sex Behaviour of the White Rat', *Genetic Psychology Monographs*, 3 (1928), 6.
3 Perry, J. S. 'Reproduction of the Wild Brown Rat (*Rattus norvegicus* Erxleben)', *Proceedings of the Zoological Society of London*, 115 (1944), 19–46.
4 Barnett, S. A. and Bathard, A. H. 'Population Dynamics of Sewer Rats', *Journal of Hygiene* (Cambridge), 51 (1953), 483–91.

Chapter 5

1 Chitty and Southern, *Control of Rats and Mice.*
2 Chitty, D. 'Rats, Mice and Agriculture', *Annals of Applied Biology*, 38 (1951), 724–5.

3 Elton, C. 'Cats in Farm Rat Control', *British Journal of Animal Behaviour*, 1 (1953), 151–5.
4 *Wild Mammals and the Land*, Bulletin No 150 Ministry of Agriculture, Fisheries & Food (London: HMSO, 1951).
5 Errington, P. L. 'Wintering of Field-Living Norway Rats in South-Central Wisconsin', *Ecology*, 16 (1935), 122–3.
6 Chapman, F. B. 'Exodus of Norway Rats from Flooded Areas', *Journal of Mammalogy*, 19 (1938), 376–7.
7 Drummond, D. C. 'The Food of *Rattus norvegicus* Berk. in an Area of Sea Wall, Saltmarsh and Mudflat', *Journal of Animal Ecology*, 29 (1960), 341–7.
8 McMurry, F. B. and Sperry, C. C. 'Food of Feral House Cats in Oklahoma', *Journal of Mammalogy*, 22 (1941), 185–90.
9 Fairley, J. S. 'The Food, Reproduction, Form, Growth and Development of the Fox *Vulpes vulpes* (L.) in North-East Ireland', *Proceedings of the Royal Irish Academy*, 69 (1970), 103–37.
10 Fairley, J. S. 'Food of Long-Eared Owls in North-East Ireland', *British Birds*, 60 (1967), 130–5.
11 Shipley, A. E. 'Rats and Their Animal Parasites', *Journal of Economic Biology*, 3 (1908), 61–83.

Chapter 6

1 Schein, M. W. and Orgain, H. 'A Preliminary Analysis of Garbage as Food for the Norway Rat', *American Journal of Tropical Medicine and Hygiene*, 2 (1953), 1117–30.
2 Davis, D. E., Emlen, J. T. and Stokes, A. W. 'Studies on Home Range in the Brown Rat', *Journal of Mammalogy*, 29 (1948), 207–25.
3 Steiniger, F. 'Revier und Aktionsraum bei der Wanderratte'.

4 Drummond, D. C., personal communication.

5 Davis, D. E. and Fales, W. T. 'The Distribution of Rats in Baltimore, Maryland', *American Journal of Hygiene*, 49 (1949), 247–54.

6 Davis, D. E. and Fales, W. T. 'The Rat Population of Baltimore, 1949', *American Journal of Hygiene*, 52 (1950), 143–6.

7 Davis, D. E. 'The Rat Population of New York, 1949', *American Journal of Hygiene*, 52 (1950), 147–52.

8 Bentley, E. W., Bathard, A. H. and Hammond, L. E. 'Some Observations on a Rat Population in a Sewer', *Annals of Applied Biology*, 43 (1955), 485–94.

9 Davis, R. A. 'Occurrence of the Black Rat in Sewers in Britain', *Nature* (London), 175 (1955), 641.

10 Schiller, E. L. 'Ecology and Health of *Rattus* at Nome, Alaska', *Journal of Mammalogy*, 37 (1956), 181–8.

Chapter 7

1 Garlough, F. E., Spencer, H. J. and Jordan, W. 'Hawaiian Rat Abatement Project. Semi-annual Report (1 Jan–30 June 1937) USDA Bureau of Biological Survey, Division Game Management Control Methods Research Laboratory, 1937).

2 Doty, R. E. 'Rat Control on Hawaiian Sugar Cane Plantations', *Hawaiian Planters Record*, 49 (1945), 71–239.

3 Schwarz, E. 'Notes on Commensal Rats', *Pests*, 11 (1943), 6.

4 Twigg, G. I. 'Infestations of the Brown Rat (*Rattus norvegicus*) in Drift Mines of the British Isles', *Journal of Hygiene* (Cambridge), 59 (1961), 271–84.

5 Twigg, G. I. 'The Distribution of the Underground Rat Population of a South Yorkshire Drift Mine', *Journal of Hygiene* (Cambridge), 60 (1962), 283–8.

Chapter 8

1 Reports to the local government board on public health and medical subjects (new series no 52) (London: HMSO, 1911).
2 Mitchell, J. A. 'Plague in South Africa: Historical Summary (up to June 1926)', *Publications of the South African Institute for Medical Research*, 3 (1927), 89.
3 Eskey, C. R. and Haas, V. H. 'Plague in the Western Part of the United States', *US Public Health Service: Public Health Bulletin*, 254 (1940), 1–83.
4 Broom, J. C. 'Leptospirosis in England and Wales', *British Medical Journal*, 2 (1951), 689–97.
5 Sharp, W. C. 'Weil's Disease in the Scottish Coal Mines', *Transactions of the Association of Industrial Medical Officers*, 2 (1953), 155–8.
6 Jenkins, T. H. and Sharp, W. C. 'Weil's Disease – Occurrence Among Workers in Welsh and Scottish Coal Mines', *British Medical Journal*, 1 (1946), 714–17.
7 Broom, J. C. 'Leptospirosis in England and Wales', *British Medical Journal*, 2 (1951), 689–97.
8 Twigg, G. I., Hughes, D. M. and McDiarmid, A. 'Leptospiral Antibodies in Dairy Cattle: Some Ecological Considerations', *Veterinary Record*, 90 (1972), 598–602.

Chapter 9

1 Chitty, D. 'A Relative Census Method for Brown Rats (*Rattus norvegicus*)', *Nature* (London), 150 (1942), 59.
2 Chitty and Southern, *Control of Rats and Mice.*
3 Doty, R. E. 'The Pre-Baited Feeding-Station Method of Rat Control', *Hawaiian Planters Record*, 42 (1938), 39–77.
4 Barnett, S. A., Bathard, A. H. and Spencer, M. M. 'Rat Populations and Control in Two English Villages', *Annals of Applied Biology*, 38 (1951), 444–63.

5 Boyle, C. M. 'Case of Apparent Resistance of *Rattus norvegicus* Berkenhout to Anticoagulant Poisons', *Nature* (London), 188 (1960), 517.
6 Drummond, D. C. 'Rats Resistant to Warfarin', *New Scientist* (23 June 1966), 771–2.
7 Lund, M. 'Resistance to Warfarin in the Common Rat', *Nature* (London), 203 (1964), 778.
8 Davis, D. E. 'The Characteristics of Global Rat Populations', *American Journal of Public Health*, 41 (1951), 158–63
9 Davis, R. A. 'Control of Rats and Mice', *Bulletin 181, Ministry of Agriculture, Fisheries and Food* (London: HMSO, 1970).

ACKNOWLEDGEMENTS

I should like to express my appreciation to many people who have helped in the preparation of this book, in particular: my friends and former colleagues in the Ministry of Agriculture, Fisheries and Food for helpful discussions; to the Controller of Her Majesty's Stationery Office for permission to reproduce plates 1, 2, 4, 5 (*bottom*), 7, 8, and Cambridge University Press for permission to reproduce fig 6; to Mr H. G. Hurrell and the Editor of the *Western Morning News* for allowing me to quote from an article by Mr Hurrell; to Dr V. K. Brown for preparing figs 1 and 2 and to Mr P. J. Cox for figs 3 and 5 and for his comments on the manuscript; finally to my wife and son for their support and encouragement.

INDEX